THE MANY FACES OF GOD

ALSO BY JEREMY CAMPBELL

The Liar's Tale

Grammatical Man

Winston Churchill's Afternoon Nap:
A Wide-Awake Inquiry into the Human
Nature of Time

The Improbable Machine: What the Upheavals
in Artificial Intelligence Research Reveal
About How the Mind Really Works

Science's 400-Year Quest

for Images of the Divine

W. W. Norton & Company

NEW YORK · LONDON

THE MANY FACES of GOD

JEREMY CAMPBELL

For information about permission to reproduce selections from this book,
write to Permissions, W. W. Norton & Company, Inc., 500 Fifth Avenue,
New York, NY 10110

Frontispiece: William Blake. *Elohim Creating Adam* (detail), ca.
1795–1805. Watercolor on paper. Photograph credit: Tate Gallery,
London/Art Resource, NY

Manufacturing by R. R. Donnelley, Bloomsburg Division
Book design by Barbara Bachman
Production manager: Julia Druskin

Library of Congress Cataloging-in-Publication Data

Campbell, Jeremy, 1931–
The many faces of God : science's 400-year quest for images of the
divine / Jeremy Campbell.—1st ed.
        p. cm.
    Includes bibliographical references and index.
    ISBN-13: 978-0-393-06179-6 (hardcover)
    ISBN-10: 0-393-06179-5 (hardcover)
1. Religion and science—History. 2. God. I. Title.
    BL245.C34 2006
    215—dc22

                                        2006017845

W. W. Norton & Company, Inc.,
500 Fifth Avenue, New York, N.Y. 10110
www.wwnorton.com

W. W. Norton & Company Ltd.,
Castle House, 75/76 Wells Street, London W1T 3QT

1 2 3 4 5 6 7 8 9 0

IN MEMORIAM

PANDORA

1937–2001

Contents

My God, my God, thou art a direct God, may I not say a literal God, a God that woulds't be understood literally and according to the plain sense of all thou sayest? but thou art also (Lord I intend it to thy glory, and let no profane misinterpreter abuse it to thy diminution), thou art a figurative and a metaphysical God too; a God in whose words there is such a height of figures, such voyages, such peregrinations to fetch remote and precious metaphores, such extensions, such spreadings, such curtains of allegories, such third heavens of hyperboles, so harmonious elocutions, so retired and so reserved expressions, so commanding persuasions, so persuading commandments, such sinews even in thy milk.

—JOHN DONNE, *Devotions Upon Emergent Occasions*

Prologue

> Theology necessarily is a system of metaphors, and doctrine represents its literalization. I am inclined to believe that the best poetry, whatever its intentions, is a kind of theology, while theology generally is bad poetry.
>
> —HAROLD BLOOM

THE POET JOHN DONNE, brilliant preacher and ironist, prayed to a God who was half literal, a bearer of plain sense, and half poetic, speaking in metaphors, "curtains of allegories," words pointing obliquely to truths that words cannot fully express. But not in words alone. Nature, in all its astounding variety, its profligacy, its beauty and cruelty, is a sort of poem. Read by a mind not irretrievably literal, and interpreted with a reverent heart, it can give a glimpse into the secret of secrets: the likeness of God. "The style of thy works," Donne said, "the phrase of thy actions, is metaphorical."

Donne was writing—and praying—in the seventeenth century, when science was making its big breakout, when the codes of nature were beginning to be deciphered and scientists of genius were making sense of a different kind of universe. And they were coming up with new, not altogether comforting metaphors to describe a different kind of God. Donne was saying that metaphors can give a glimpse of the ineffable, and we can look for them not only in scripture but in nature, in the physical world.

Devotions Upon Emergent Occasions, some lines of which appear as an overture to this book, was written in 1624. In that same year Galileo, another author of wit and bold imagination, began to write the book that would bring him into sharp and painful collision with the Catholic Church. That book was *Dialogue Concerning the Two Chief World Systems*, which subtly—though not quite subtly enough—argued the Copernican case, that the Earth moves round the Sun, not vice versa.

Donne and Galileo. One praised as a "Copernicus in poetry" for his daring innovations in verse. The other a promoter of a scientific theory deemed a threat to faith and doctrine by the princes of the Roman Church.

Two winters after Donne died, famously dressing himself first in his shroud so an artist could capture his likeness, Galileo was brought to book. He was summoned to the Holy Office at the Vatican in Rome, accused of teaching heresy. His clever literary maneuver of putting his case for Copernicus in the form of a discussion among friends with contrary opinions had backfired. Galileo disingenuously claimed the book showed he actually opposed Copernicus. His interrogators, no simpletons, suspected they were being taken for "a pack of fools." Galileo was brought into a room next to the Church of Santa Maria Sopra Minerva, now part of the Italian Parliament. There he was told to kneel down and formally abjure his theories as "errors and heresies" that he now despised. As part of his punishment, he was ordered to recite the seven penitential psalms once a week for three years. The next day he was sentenced to house arrest at his villa in Florence.

Two men of genius on their knees. Donne in a posture of prayer to a God he regarded as a mystery, a deity who spoke in poetry as well as prose, whose attributes were paradoxical. God to Donne: poet to poet, neither committing himself to irretrievable, plain statements of fact. "He that seeks proofe for every mystery of Religion shall meet with much darknesse," Donne had warned. Prayer, for Donne, was a sort of wrestling match with an opponent whose intentions were never quite clear. It was a way to "batter heaven," an approach that sometimes verged on violence, the violence of a supplicant not certain his petitions will be granted. "Prayer," said Donne, "hath the nature of Impudency."

And then Galileo, kneeling at the command of his judges, no martyr choking on smoke and flames but in mortifying surrender, denying that a belief he held with the utmost conviction was true. Galileo was beyond the protection of literature. He had made his opinions too evident. His offense was to treat the Copernican theory not as a mere artifice of mathematics but as a literal, physical truth. Galileo was saying in effect that there is only one way in which the heavenly bodies are organized, the Copernican way. The Church by contrast held that God can move the stars and planets in any way he chooses, any time he chooses. To say otherwise is to cast aspersions on God's sovereign power. And that, in the view of Rome, involved a heresy. It altered our conception of God by casting aspersions on his omnipotence, a story we shall investigate in detail later on.

Donne, on the other hand, gave no offense to the authorities because he declined to say that *anything* was the literal, physical truth. He left God much as he found him. Donne was brought up a Catholic at a time when Catholics were subjected to harsh persecution. He was taken by his boyhood tutor to watch Catholics being hanged, in order to intensify his anti-Protestant ardor. But he made the transition to the Protestant faith discreetly, sometimes ambiguously, behind a screen of paradox and wit, and of early poems more opaque than revealing. The doctrine of the Trinity, the cornerstone of his religious imagination, he proclaimed to be an inscrutable riddle, as baffling as any in the whole of Christianity. There is not, he said, "so steepy a place to clamber up, nor so slippery a place to fall upon," as the idea of three persons in one God. It is not a theorem in mathematics. Donne made God more of a mystery than before, and kept on the right side of ecclesiastical authority. Galileo made God a fraction less of a mystery, and suffered for his pains.

Much has been written about the decay of belief and the "death of God" during the modern period. But there has been less discussion of how science has, over the past four centuries, reinvented God to suit its own shifting view of the world physical and the world biological. God has many faces, and modern theoreticians, who have shown nature to be more subtle than anyone in the seventeenth century could have supposed, have not been shy of translating nature's sophistications into new metaphors for nature's Author. Since Galileo, God has been given multiple aspects and contradictory portraits by scientists, some of whom have not finally made up their minds as to what sort of universe they want and are in some uncertainty as to what kind of God they can imagine.

It was in the seventeenth century, when breakaway religious sects and quaint heresies proliferated, that science explicitly offered itself as a mediator between the human and the divine. Isaac Newton suggested this lofty role for science, as did the newly established Royal Society. The Reformation, acting with brutal thoroughness, had dismantled most of the mediators helping to bring the ordinary person closer to the mystery of the divine: angels, prophets, saints, martyrs, relics, shrines. The absence of these third-party intermediaries left a vacancy that the new physics and astronomy aspired to occupy. Newton's contemporary Robert Boyle, founder of modern chemistry, believed he and his peers had a privileged connection to the Creator, that the scientist was "born a priest of nature" and that his researches "mediate between God and the creation." In Boyle's mind the scientist was to some extent making, or remaking, God in science's image. He was domesticating the divine. No longer was it impolite to ask what sort of God God is, where we should put him, and what we can find for him to do.

In Newton's day, theology dealt with the mysterious and strange, science with the intelligible and provable. Newton's universe was an efficient piece of clockwork, all the parts working together harmoniously, though there is evidence that Newton himself was not quite satisfied with such a humdrum state of affairs. He did recognize how powerfully the human mind is attracted to mystery; people are inclined to regard as holy what is least understood. Such a notion of God, he said, is hazardous and can easily lead to a denial of his existence.

Today, the situation is somewhat different. It is modern physics that offers strangeness and indeterminacy, while theology in certain of its idioms is criticized for not being strange enough, for shying away from the unlikely and implausible. It is taken to task for being too timid and avoiding the *un*harmonious and dissonant aspects of existence. John Polkinghorne, a particle physicist who took holy orders in the Anglican Church, thinks that while the imagination of scientists has grown bolder and more daring, that of theologians remains parochial, human-centered, and too reluctant to shock and overturn common sense. Einstein was wrong, Polkinghorne thinks, in confessing an amazed reverence for the harmony of natural law, which was Einstein's idea of God. That does not begin to describe the elements of risk, terror, and amazement that are present when the human comes into contact with the divine.

The English botanist John Rodwell also thinks today's religious establishment turns a blind eye to the exciting ways in which science, especially physics, is coining new metaphors for God, preferring to focus instead on images of love and caring that are immune to intellectual fashions. "I'm not very optimistic about the church's ability to take the world seriously," Rodwell said in an interview, "and that would include the natural world." The cognitive scientist Brian Cantwell Smith insists that the kind of metaphors used to describe God *matter*. They are not mere poetry. Bad, misleading ones can split science and religion into opposing factions. Religious authorities, Smith says, must recognize that their metaphors for God are not working as well as they used to, not only for people outside the community of believers but for those inside as well. He goes on: "This is especially true for conceptions of 'God' as somehow *separated* from the world. In fact, I'd argue that the idea of a 'separated' God no longer makes any sense at all, in the context of enmeshed, modern science. It seems like a downright dangerous idea."

Newton gave us a God of awesome power, sovereign, absolute. He was Lord of Dominion, ruler supreme. At the same time he was given specific, somewhat menial tasks to perform, among them making periodic corrections to the motions of the celestial bodies. The Church made the mistake of accepting this role for God. When it became clear that the heavenly bodies did not need divine intervention to keep them on an even keel, the Church was left stranded and defenseless against the arguments of atheists, that a God with nothing to do is no God at all.

Modern physics provides us with dramatic evidence of ambiguities, cloudy regions of uncertainty at the heart of things, gateways where an active God might enter discreetly and modify the course of events without anyone noticing. Such patches of haziness are located in the depths of the quantum microworld and the large-scale unpredictability of chaos theory, where virtually identical causes can lead to wildly dissimilar results. Curiously, however, investigations of such a possibility tend to posit a God subtle, discreet, and hidden, a self-effacing God who is under the constraint of physical law in a way that would have appalled seventeenth-century believers in an omnipotent, willful deity.

The trouble is that there is no single, all-inclusive portrait of God that suffices in all circumstances. And often it seems that people, even when open to the idea of multiple deities, do not find it easy to worship more than one God at a time. Newton himself met this dilemma. Newton placed an almost obsessive emphasis on God's total dominion. He did also write of God's love and care for his creatures, but these sentiments are usually in the form of a brief tailpiece or postscript to a statement that gives priority to God's overriding potency and will. He was more dubious about the reliability of the texts of the New Testament than about those of the Old Testament. And Newton's picture of God as the overseer of order, not of confusion, gave a misleading impression of a well-behaved and "caring" universe. Today, the image of a reasonable, circumspect, and tactful divinity who hides behind the veil of quantum or chaotic uncertainty does not begin to do justice to the Old Testament God of wrath and vengeance who transcends our notions of morals and decent conduct. Polkinghorne's God is a lover, a sharer, a friend. But it was the Danish thinker Søren Kierkegaard who said that we cannot know God as a friend until we have first known him as an enemy. He has to break us down before he can build us up.

Doubtless science will continue to mediate multiple images of the divine. People want new and exciting, "improbable" theories of nature, and they seem to be open to a variety of profiles of God. Opinion surveys confirm an upsurge in "spirituality" as opposed to "faith," a hands-off culture of religious experiment where doctrine and creeds are superfluous and personal experience is all. Analysts of this trend speak of "God Lite" and "cafeteria religion" to describe a privatizing of religion and a sense that "I decide what God is." There is an indifference to theology.

Science's models of the divine are subject to obsolescence. Newton and his followers have been accused by the modern historian William Placher of presenting a deity overbearing, a willful ruler of a universe with no will of its own, a being locatable in specific places and not in others. This image, Placher says, is responsible for the shallow and trite images of God in circulation today. But today's science is not a trite activity. Its insights into the nature of divinity may not be the whole story by a long chalk, but that is the way mediators have always been. It was Job's misfortune that he had no mediators, good or bad, to make some sense of a God of baffling otherness, to ease the terrifying ordeal of unprotected contact with the divine by rendering him in quasi-human terms.

Theology might do well to consider what appears to be a double challenge to its traditional role as a source of God knowledge. On the one hand, the new spiritual experimenters seem apathetic to theology; they are seeking their own path in their own way. On the other, a priest-scientist like John Polkinghorne is not a lone voice when he suggests that science is stealing theology's thunder, that its imagination—and John Donne knew that theology is in large part a work of the imagination—is racing ahead, leaving theology limping along behind. For theology to assert itself in this contest, however, may not be a wholly popular move. The spiritualists assume that the vehicle of religion they choose determines what God is. The Church, by contrast, takes exactly the opposite view. It teaches that God determines what faith is. Such a view is a restriction on the "many faces" of God it is permissible to invent. One option is for theology to compete with science, though in its own incomparable fashion, preserving its unique character, its sense of mystery, and the huge paradox of one God presiding over a world of many faces, in which joy and misery, good and evil, justice and inequality, health and disease, co-exist. The other option, suggested by a scholar who has studied the dangerous, opaque, even "sinister" God of the Book of Job, is to decide that none of the ingenious images of the divine proposed by modern physics can explain the brute facts of human existence, and that between the two sides, religion and science, there should be an amicable separation.

THE MANY FACES OF GOD

CHAPTER ONE

Smashing the Idols

And so it will be until the end of the world, even when all gods have disappeared from the earth: they will still fall down before idols.

—FYODOR DOSTOEVSKY

This idea that the human mind is a perpetual manufacturer of idols is one of the most profound things which can be said about our thinking of God. Even orthodox theology is often nothing more than idolatry.

—PAUL TILLICH

THE SEVENTEENTH CENTURY, birthtime of modern science, was also witness to the final convulsions of a movement intended to crush superstition and bring the ordinary person into a more wholesome relation to the divine. This was the era of iconoclasm, the wholesale destruction of sacred images, paintings of holy beings, stained glass apostles, statues of the Virgin, carved likenesses of saints. To the ordinary churchgoer these were aids to devotion. Popular art provided the "sweet deceits" of popular piety. But to the Protestant Reformers they were idols, to be not simply removed from churches but smashed, annihilated, as Moses, in a fury, shattered the Golden Calf of the Israelites to smithereens.

John Calvin considered such images idolatry; any attempt to render the invisible visible was "an impious lie." He likened idolators to lavatory attendants. "Hardened by habit, they sit in their own excrement, and yet believe they are surrounded by roses." The paintings and statues in churches pandered to a "foolish and inconsiderate longing." What made Calvin take them so seriously as a menace to true religion was that the longing, foolish though it might be, was so deeply embedded in the human psyche. Human nature seems to have an unquenchable need for images of the divine, and a distinct uneasiness about confronting God alone. "Daily experience shows," Calvin said, "that the flesh is always restless until it has obtained some figment like itself, with which it may vainly solace itself as a representation of God."

The veneration of images, the kissing of plaster feet, the licking of wooden faces of saints in their niches, the lighting of candles, praying to patron saints of cattle and hogs, were swept away by the iconoclasts. But that left an empty space which, given the popular bent for the tangible, invited other means of filling it. Christopher Hill, a historian who may almost be said to have reinvented our understanding of the seventeenth century, noted how persistent was the popular notion of mediators, bridges between the supernatural and the ordinary, even at a time when city life was growing more sophisticated and heresies abounded. Hill described how the vacancy was partly satisfied by science as it was developed and given a religious dimension by the early Royal Society. The harmony and precision of the Newtonian clockwork universe, presented as the work of an intelligent Creator, helped to compensate for the loss of contact with a higher and holier reality than had been offered by the toy divinities a culture had venerated, then smashed to pieces.

Since the Reformation, a war against "idols" had been gathering momentum, violently spasmodic and hinting at forces of social unrest using iconoclasm as a vehicle for protest. Early one morning in August 1566, an English financial agent named Richard Clough strode reeling out of the cathedral at Antwerp, in shock at the spectacle of utter desecration he had witnessed there. "Coming into the Church of Our Lady," he reported, "it looked like a hell, where were above 10,000 torches burning, and such a noise as if heaven and earth had got together, with falling of images and beating down of costly work: such sort that the spoil was so great that a man could not well pass through the church." Clough told his employer, the financier Thomas Gresham: "I cannot write you in ten sheets of paper the strange sight I saw here, organs and all destroyed."

It was the sheer mob fury, the nearly ecstatic rage for demolition, that sent the young Englishman into the street in a state of dazed incredulity. But on a less grandiose scale the forcible removal of sacred images, religious paintings of priceless value, church ornaments, irre-placeable sculptures, had been going on for several years. Martin Luther had taken a rather restrained view of religious art; but he was no match for a more radical colleague at Wittenberg, Andreas Karlstadt, who made a brutal assault against all paintings depicting supernatural beings, as well as the artists who executed them. The Swiss religious leader Huldrych Zwingli, though an admirer of the Renaissance humanists, was incensed by the decadence of the Roman Catholic Church and ordered that all churches be stripped of their works of religious art.

The painter Holbein had fled the city of Basel thirty-four years ear-lier and returned to England, chilled by the ferocity of the Protestant campaign against sacred images. During the 1530s Basel was a major cul-tural center, home of the humanist thinker Erasmus. Unchecked by an inactive and ambivalent magistracy, popular emotion boiled over and the mob, in Erasmus's words, started to "commit frightful ravages." Hundreds marched to the cathedral, broke down the doors, and smashed every cult object in sight. The debris from all this destruction was taken out and thrown upon bonfires, while the other Catholic churches were ransacked. It was only thanks to the alert intervention of the town sec-retary, Johannes Gerster, that Holbein's painting *Solothurn Madonna* sur-vived the vandalism of the reformers and is with us today. In the end, the episode was a political coup as well as a religious one. The authorities were somewhat abashed at such a show of mob force. A dozen Catholic magistrates resigned and the guilds were granted a greater say in the running of the city. Guilds were collections of lay people, sometimes for the purposes of commerce, but usually to provide a center for religious activities, especially prayer, for their members, outside the control of the clergy, who were paid to say mass. They reflected the gap between Church professionals and popular piety; some of their practices, espe-cially flagellation, were disturbing to the authorities.

England, though facing its own religious crisis, was a refuge for Holbein. The forcible removal of images and ornaments was happening there as well, but in a less violent and more hesitant fashion. In the 1520s incidents of iconoclasm had occurred, but they were mostly seen as acts of ruffians on the fringe of Protestantism. When a gang of incendiarists broke into a church at Rickmansworth, Hertfordshire, at night and set ornaments, paintings, and the organ ablaze, it was thought to be the work of heretics. The congregation agreed. Cardinal Wolsey was able to finance a restoration of the nave by offering remission of 140 days in purgatory to anyone who contributed money.

At the time, English parish churches were crowded with images of saints, martyrs, the Virgin Mary, the Holy Family, many in niches hung with curtains. A movement called the Cult of the Holy Name, sponsored by Lady Margaret Beaufort, encouraged the setting up of altars to Jesus and celebration of the "Jesus Mass" in parish churches. These innovations marked the care and closeness of Christ in his human aspect; they helped to make him a more approachable figure than the undepictable, remote God the Father in his glory. But in the 1530s, the atmosphere changed. New clerics came in having scant sympathy for the comfort people took from images of the Virgin, sometimes pictured in tears, and the tenderness of the Jesus cult, the sense of nearness to the divine that came from touching and kissing images. They pressed parish priests to stamp out superstition and beliefs that had no basis in the Bible. In 1537, the new dean of Exeter Cathedral, Dr. Simon Heynes, wrote a memorandum in which he complained that the "common people" had a greater trust in outward rites and ceremonies than was decent, and were apt to value above the teachings of scripture such dubious distractions as "images and adorning of them, kissing their feet or offering candles unto them." The craven clergy, Heynes said, fearing loss of income if such practices were stopped, were egging on their parishioners to such idolatry instead of directing them to more austere forms of piety.

Some churches hid their treasures; others sold them to private collectors. The destruction was not on the ferocious scale of Antwerp and Basel, but still, frescoes were blotted out with whitewash, stained glass windows shattered, memorials defaced. The wreckers mockingly degraded the ransacked art, in one instance using hollow statues of the

Virgin as a pig's feeding trough. John Donne, who especially prized the visual sense, stood out as a defender of images. In a sermon preached at St. Paul's Cross on May 6, 1627, he wished woe to "peremptory abhorrers of Pictures, and to such uncharitable condemners of all those who admit any use of them, as had rather throw down a Church, than let a Picture stand."

Iconoclasm surged in England during the Civil War of the 1640s and the social upheavals that went along with regicide and the turmoil of the republican experiment. Even before the Civil War, idol-smashing was winked at by politicians trying to curry favor with the working class, whose militant elements acted out their class antagonism by defacing the opulent tombs of the nobility and smashing stained glass windows embellished with the coats of arms of the rich and titled. When they could afford to dispense with such support, the politicians tried to stamp out iconoclasm.

Oliver Cromwell, the parliamentary army's leading general in the Civil War, later becoming Lord Protector of the realm, was a sincere adherent of the teachings of Calvin. He had espoused a particularly intense brand of Protestantism while recovering from a species of nervous breakdown. In later years he turned sharply against church vandalism, but during the war he acquired a black reputation, only partly deserved, as an iconoclast. Royalist propaganda sheets accused him of defacing Peterborough Cathedral, where it was said his soldiers wrecked the pulpit, played lewd tunes on the organ, and danced up and down the nave in surplices. Cromwell was reported to have wrenched out a crucifix himself and allowed his men to stable their horses in a profaned Lincoln Cathedral. In the spring of 1644 at Cambridge, he allegedly approved tearing up the Book of Common Prayer in the faces of the university clergy.

At Ely Cathedral in January 1644, Cromwell ordered choir services stopped. They were to be replaced by edifying Bible readings and sermons. When his order was ignored, he arrived at the church with his hat on while the choir was in full voice and commanded the service to cease at once. No notice was taken of his outburst. Cromwell returned in a rage, herding the entire congregation out of the cathedral, his hand on his sword. But he and his troops were not the only wreckers of holy places. His Royalist opponents were less than models of restraint when it came to such desecration. Montrose stabled his horses in a church in Scotland; other Royalist officers stored gunpowder and cannon balls in parish churches. Pigs and horses were baptized and ecclesiastical documents put to the flames.

Some destruction was aimed at disabusing the unthinking masses of the superstition that any middling agent, except Christ himself, was needed to connect the realms of heaven and earth. The pope was highly dispensable in this role. When Sir William Brereton, a Puritan magistrate and Civil War activist on the parliamentary side, blasted to rubble the spire of Lichfield Cathedral with his artillery, he waved aside objections by saying it had reminded him of "the Pope's Triple Crowne."

The poet Milton—whose God in *Paradise Lost* is quite impersonal, his warmth almost never rising above room temperature—thought the sentimental images of Mary, all tenderness and tears, marred the original purity of Christianity and compromised the singleness of God's majesty and power. He called them "gaudy glistenings," reminding him of the "hellish sophistry of papism." There was a snobbish element in much of this. Idol worship was for ignorant peasants, strangers to theology, a harking back to the crude magic of the Middle Ages at a time when society was becoming more urban and literate. But the popularity of icons of one sort or another was not confined to the unlettered. A notorious book, the *Eikon Basilike, the Portraicture of His Sacred Majestie in His Solitude and Sufferings*, was published early in 1649, a few days after the execution of King Charles I, and was a runaway popular success. Written by the Reverend John Gauden, a Presbyterian clergyman, perhaps based on writings by Charles himself, the book idolized the dead king and made him a holy martyr, of the order of Christ himself. Gauden's highly colored prose style, a seventeenth-century version of modern tabloid journalism, appealed widely to the masses.

A different kind of iconoclasm was carried out by Sir Francis Bacon, midwife of the new science and dismantler of obsolete superstitions. Bacon knew science in its modern form could not be brought to birth successfully without the drastic exposure and removal of certain built-in or acquired mental biases, deeply embedded myths and plausible illusions. The mind, he thought, is apt to "leap and fly" from particulars to universal concepts we mistake for infallible and unalterable truths. It was no wonder that the science of his time was stuck in a rut. A very human tendency to adhere to prior theories and beliefs as if to Holy Writ, ignoring evidence to the contrary, meant that science simply spun its wheels and could not move forward. Such mental attitudes Bacon called "idols," counterparts to the false gods and painted emblems the Protestant Reformers rejected as an impediment to understanding what religion is really all about. These preconceived notions were supposed to mediate between the human mind and reality, but in fact they were a diversion, sending it off in false directions. There are fascinating echoes of Calvin in Bacon's writings on this topic, especially the idea that idolatry is in part a clinging to the past and also a surrender to the inclination of flawed human beings to go with the crowd.

Any impediment blocking the advance of the new science Bacon regarded as a kind of heresy, inviting a comparison with religious heterodoxy, which ran rampant in the aftermath of the Reformation. By using the word "idol," he aligned himself intellectually with the iconoclasm of the church wreckers. Whereas Protestant authorities banned the veneration of sacred images giving the illusion of nearness to the divine, Bacon for his part deplored the "veneration of empty ideas." In both cases, artifacts supposed to point to a genuine reality only succeeded in becoming objects of worship themselves. Idolatrous leanings of whatever kind "need to be smashed, neutralized by those who have shattered idolatrous tendencies in themselves." As Bacon wrote: "For let men please themselves as they will in admiring and almost adoring the human mind, this is certain, that as an uneven mirror distorts the rays of objects according to its own figure and section, so the mind, when it receives impressions of objects through the sense, cannot be trusted to report them truly, but in forming its notions mixes up its own nature with the nature of things."

Perhaps this was too bleak a view. Bacon was of a radical mind, bent on overturning centuries of fixed ideas, ingrained attitudes, and dogged myths, chasing them out of the temple of science. He wanted to cleanse the human mind of its preconceived notions, much as the Puritan ransackers rudely reduced church interiors to bare whitewashed walls. The question is, however, whether either religion or science can exist without the aid of mediators, who assist in the difficult task of connecting different realms of being. Science *needs* a mind that is not entirely naked, not stripped bare of its theories, thematic ideas, and points of view.

Mental biases, by closing a scientist's mind to certain possibilities, open it to other, more daring and original ones, or so the history of science suggests. Einstein was influenced in his thinking by a whole collection of concepts that were not strictly scientific, but partly aesthetic, such as that nature is not unreasonably complex, and that the universe is essentially harmonious and intelligible to humans. Ideas about symmetry and simplicity were also powerful considerations in Einstein's work, leading him to expect certain things of nature and to rule out others.

In the seventeenth century, the role of mediators, like much else, was in flux. What had once been fairly comfortable now became awkward, problematic, a thorny question. How was the rift between the world material and the world spiritual to be spanned? The Reformation had abolished mediating saints and discredited the authority of priests, as well as the infallible quasi-divinity of the pope. But other elements came in as substitutes. The devil emerged as a vivid and strong personality. Characters like John Knox claimed to be latter-day prophets, mediators who needed no mediator of their own, since God now spoke directly to them through the individual conscience. Astrologers did a good business. During the time of the English Civil War, under the rule of Cromwell and then the restored monarchy, popular taste favored interpreters, whether of the stars and their influence on daily life, or of scripture or folk myths. Christopher Hill, who saw a "crisis of mediation" emerging in the seventeenth century, writes of the stress experienced by Protestants who were now on their own, making contact with the divine as best they could, missing the emblems that interceded for them in pre-Reformation days. "Men emancipated themselves from priests, but not from the terrors of sin, from the priest internalized in their own consciences," Hill comments. "Only very strong characters, or the very fortunate, could stand the strain. Unmodified, it was a doctrine more appropriate to crisis conditions of struggle than to normal living in a stable society." It might actually foster instability. The Protestant rule was that there are sheep and there are goats, the elect and the unsaved mass. Confession, absolution, were obsolete practices, which left the problem of what to do with people in doubt as to their salvation.

The iconoclasts were not haters of art per se, but they were intensely hostile to any form of mediation they felt was false and hindered access to the truth. In late medieval times, there was an immense desire to draw closer to the divine. Lay persons had become a significant force in the body of the Church, and their fervor produced a good deal of tension, as daily life became saturated with mediators of God and religion. Great efforts were made to capture the Absolute in material, tangible form and embody the ineffable in shapes one could see and touch. At its crudest, this was a way of exploiting supernatural power. For a lay person, a religious image might affect the vast operations of nature, perhaps calming whirlwinds or alleviating drought. The Church did not always discourage such superstitions, by means of which huge, inscrutable forces were domesticated. By the fifteenth century there had arisen, according to the American scholar Carlos Eire, a devotion to the saints so overarching it amounted to a kind of polytheism. They crowded the pages of popular religious literature and were the topic of sermons. They offered an indirect means of access to God. A particular saint looked after specific sections of the laity and took care of limited kinds of emergencies. A saint could cure a certain disease, or inflict it on someone who deserved it. Professor Eire writes that late medieval Catholicism "developed a nebulous concept of divinity: God and the saints became an indistinguishable source of power and wrath." People were uncertain as to whether they were saved, and yearned to be saved. By domesticating the mediators between God and humankind, late medieval society aimed to compel a guarantee of salvation. Ironically, this introduced a new source of anxiety: the mediators tended to be seen as godlike anyway, and had to be mollified in case they developed a godly anger. The dividing line between the divine and the human began to blur. There were few precautions in place to avoid confusing one with the other. Even mass, increasingly a separate ritual, flamboyant and replete with imagery, became another material realization of the supernatural, channeling power to the otherwise powerless.

The iconoclastic Protestant movement swung in the opposite direction, toward a strict, uncompromising doctrine of God as totally "Other." The divine was removed to its own realm, separate from the created world. According to this group, God's power was absolute, and the rules by which he operates could not be altered willy-nilly, as was the case when lesser supernatural beings acted as divine agents in charge of specific aspects of daily life. A wedge was driven between the spiritual and the material. This could not help but have repercussions outside the religious arena. Smashing idols was an affront not only to the Catholic Church but also to the secular authorities. It set the scene for social upheavals that had more to do with the abuse of worldly than divine power. A revulsion against everything regarded as false in the relation between God and humans led to a redefinition of just what that relation ought to be, and, inevitably, what kind of God we have. It also posed a stark threat to governments whose relation with the governed might be considered false and in need of redefining.

The campaign against idolatry changed religion in ways that were not immediately apparent at the time. Removing the prolific images of the Virgin Mary from church walls helped to make devotion more of a masculine affair. A maternal icon was replaced with a paternal one, and a filial one: Father and Son. Piety was defeminized. The God of Newton, a lifelong bachelor and perhaps a virgin to the end, is intensely male, with power his primary attribute. Carlos Eire thinks iconoclasm also transformed religion from an experience of aesthetic and largely visual aids to devotion into something much more austere and intellectual, mediated by language. And in some respects it moved from exterior to interior, intangible and unseen, needing the supervision of a clerical elite. "In practical terms," Eire says, "this meant a very concrete shift from a world brimming over with physical, visible symbols that were open to a rather wide range of interpretations—some intentionally ambiguous—to one charged principally with verbal symbols that were subject to the interpretation of a carefully trained and ostensibly learned ministry."

A new code of control emerged in the reformed Church that placed tight curbs on the way scripture was to be interpreted, by an autocratic hierarchy, leading critics to dub the reformed clerisy "the new papists." As the Bible became the one touchstone of faith, not only the interpreters needed to be more literate; so did their lay audiences. In the role of educators, the priestly class exercised even more control. If the power of images was replaced by the power of words, piety could be regulated more stringently because language, when studiedly prosaic, avoids the vagueness, the emotional license, afforded by paintings and wooden likenesses of the divine. "Those who broke the law by overturning altars, smashing images and feeding the host to animals," Eire notes, "eventually came, wherever they were truly successful in their revolt, under the control of a consistory."

It may be no accident that today some religious thinkers with an ecumenical bent are talking about a "theology of mediation." By this they mean that one way of making common cause with a range of faiths is to see God's relation to the world as operating via a number of agents—not necessarily an incarnation of the divine, as Jesus is believed to have been, but mediators with a variety of roles, communicating different aspects of a God who is far less simple than the learned doctors of medieval theology supposed. These mediators might be said to present different Gods. For Calvin, such a system would have been a heresy tantamount to polytheism, to idolatry. For us, looking back at the troubled history of a piety ruthlessly stripped of its material connections to the unknown, the idea of multiple mediators, highlighting a complexity foreign to the narrow compass of the Reformers, might make a good deal of sense.

CHAPTER TWO

God's Immensity, Man's Dilemma

The universe needs one creator with
a taste for opposites.
—TERTULLIAN OF CARTHAGE

It is narrated of the biologist Haeckel that some one
once put to him the following provocative question:
"If in some way you could address a question to the
Universe and be assured of a truthful answer, what
question would you ask?" Haeckel remained silent for
some moments absorbed in thought, and then he said,
"The question I would most like to see answered is this:
Is the Universe friendly?"
—JAMES H. BREASTED

O NE NEW FACTOR added to the intense focus on God as super-power as the modern age got underway. There was an explosion in the imagined size of the universe. In place of the human-centric medieval cosmos tightly bounded by the sphere of the fixed stars, a place of comfortable dimensions, the idea took hold in the sixteenth and seventeenth centuries that the universe was infinitely large and, what is more, that the stars were really suns similar to ours, surrounded, as ours is, by planets, and these planets inhabited by rational creatures. That did not necessarily follow from the revolutionary theories of Copernicus. A glimmering of this notion occurs in St. Augustine, who said the universe must be infinite because God's omnipotence requires it. To say that God is enclosed in a tiny space makes no sense, Augustine thought, though he considered speculation about other inhabited worlds a pagan error that libeled God's wisdom. During the prelude to the scientific revolution of the seventeenth century, arguments for the existence of other worlds relied extensively on the unrebuttable premise that God's majesty required it.

Thomas Digges, a sixteenth-century thinker who was the first translator of Copernicus into English, tilted at the human presumption that we can see all there is to see in the universe in his book *A Perfect Description of the Celestiall Orbes,* published in 1576. In fact, said Digges, there must be innumerable stars one cannot see with the naked eye. These invisible bodies "may be thought of us to be the glorious court of that great God, whose unsearchable works invisible we may partly, by these his visible, conjecture." Giordano Bruno, born in 1548, who sometimes made a thorough nuisance of himself while promoting new ideas, including a theory of space that partly anticipated Newton's, held that there must be other worlds like ours since God's "infinite excellence is incomparably better expressed in things innumerable than in things merely numerable." In the early modern period the polymath John Wilkins, bishop of Chester and secretary of the Royal Society, flatly stated the axiom: the more worlds, the more glory for their creator. In his provocative book *The Discovery of a World in the Moone* (1638), Wilkins predicted that in the fullness of time, men would land on the Moon and find it much like Earth. He was clearly influenced by the voyages of discovery that were then expanding people's idea of the world. "We have not now any Drake or Columbus to undertake this voyage," Wilkins wrote, "or any Daedalus to invent a conveyance through the air. However, I doubt not that time who is the father of new truths, and hath revealed unto us many things which our ancestors were ignorant of will also manifest to our posterity that which we now desire but cannot know."

Blaise Pascal famously said the vastness and silences of interstellar space scared him, but others began to find the earthly globe a little constricting. In one of the most celebrated books of popular science circulating during the seventeenth century, *Conversations on the Plurality of Worlds,* Bernard le Bovier de Fontenelle confessed that he had been feeling a little cramped in a universe limited to just what he could observe. The scene is an elegant supper in a spacious park, the moon high in a cloudless sky. His partner in conversation is a witty and beautiful marchioness. "When the sky was only this blue vault, with the stars nailed to it, the universe seemed small and narrow to me; I felt oppressed by it," he confides to his fair companion. "But now that they've given infinitely greater breadth and depth to this vault by dividing it into thousands and thousands of vortices, it seems to me that I breathe more freely, that I'm in a larger air, and certainly the universe has a completely different magnificence."

Isaac Newton himself wrote in Query 31 of his *Opticks* that "it may be allow'd that God is able to create Particles of Matter of several Sizes and Figures, and in several proportions to Space, and perhaps of different Densities and forces, and thereby to vary the laws of Nature, and make Worlds of several sorts in several Parts of the Universe. At least, I see nothing of Contradiction in all this." And the idea was given a royal airing. In a sermon preached to Charles I, John Donne said that "subtle men, with some appearance of probability, have imagined many worlds as large as our own."

The many worlds hypothesis and its subtle promoters presented the Church with a dilemma. On the one hand, it was subversive and radical. It made mock of the biblical Creation story and was quite unscriptural. But to deny it flat out meant undercutting the crucial Church doctrine, amounting almost to an obsession, of God's absolute power to create in any way he chooses. At the same time, it invited the conjecture that the universe was not only bigger and more tremendous but also more impersonal than had been once imagined. The idea arose of an *uncaring* universe, indifferent to the fate of earthlings, in which humankind is newly set adrift. In his letter to the Grand Duchess Christina of Lorraine, Galileo had stressed that nature is quite unconcerned as to whether or not humans can make sense of what she is up to. Nature is "inexorable and immutable; she never transgresses the laws imposed on her, or cares a whit whether her abstruse reasons and methods of operation are understandable to men."

Such a disconcerting vision of the universe naturally affected people's picture of God. The king of an infinite universe must be infinite in scale. Often, in the Hebrew and Christian scriptures, the image of God as somewhere else, aloof, uncommunicative, turning his back on his creation and his creatures, emerges when the idea of his nearness is at odds with the immensity of his powers. And when God becomes less available, mediators are in greater demand and so become more plentiful.

Throughout history, there were marked shifts in how people viewed this particular aspect of the divinity, God's closeness to or distance from us humans. Early on in the Bible story, God makes himself accessible. After Adam eats the fateful fruit of the tree of the knowledge of good and evil, God walks in the Garden of Eden in the cool of the day looking and calling out for him. Later on, he arranges a rendezvous with Moses, and appears to Abraham and Jacob. But midway through the story, something happens. God is still omnipresent, all-knowing. But he is less approachable than he once was. He no longer makes the first overture, nor does he go out of his way to ease the task of finding him, a task now becoming difficult, even something of an ordeal. It might also call for prayer and abstinence, some discomfort and privation.

In the Hebrew scriptures, the *tanakh*, God stops speaking altogether after the Book of Job. Jack Miles, author of a literary treatment of the Old Testament, thinks it highly paradoxical that, while the Bible leads the reader to suppose that in the Book of Job God ends the encounter with Job by putting Job firmly in his place, reducing him to silence, it is actually Job who seems to have rendered God speechless. When Job pleads for answers to the question of why he is being made to suffer in this egregious fashion, God seems to talk right past him, his words patently not to the point. They are all about God's awesome powers of controlling nature. By speaking out of a whirlwind, he gives the impression of an overwhelming natural force rather than a being interested in the welfare of his creatures. It is God the omnipotent superintendent talking, his power confirming his right to give irrelevant answers to inquiries which for Job are matters of supreme urgency. "Have you an arm like God?" Job is asked. And, "Can you thunder with a voice like this?"

"Divine neglect" is how the Hebrew scholar James Kugel describes the growing elusiveness of the deity as chronicled in the Bible narrative. This strange dwindling of interest coincides with the reduced status of the Old Testament prophets after the death of Malachi, the polemical anti-divorce preacher who warned people that Yahweh would come in judgment when they least expected it. As the biblical period moves forward, God's aspect undergoes a double alteration. He becomes bigger, and at the same time grows increasingly distant and remote. In the Book of Isaiah, God expands to giant stature by comparison with the slight and paltry dimensions of the mundane world. In Isaiah 40:15–17 we are told:

> Behold, the nations are as a drop of a bucket, and are counted as the small dust of the balance: behold, he maketh up the isles as a very little thing.
>   All nations before him are as nothing: and they are counted to him less than nothing, and vanity.

In Kugel's words, "The same God who buttonholes the patriarchs and speaks to Moses face to face, is perceived in later times as a huge, cosmic deity—not necessarily invisible or lacking a body, but so huge as to surpass our own capacities of apprehension, almost our imagination."

Another explanation for the altered image of God in the Bible—larger in size, more distant in aspect—was the growth of new thinking about monotheism. If there is only one God, it stands to reason that he cannot be confined to one locale, one people, even if that people, that place, are part of his design and strategy for the future of the world. He must be a God for all places, knowing everything, being in charge of history. He is not someone you would expect to meet by chance on a street corner, who has time on his hands to chat.

"That is the God, or close to the God," Kugel affirms, "of later Judaism and Christianity—ungraspably big and far off, who rules the whole world and 'calls off the generations' from His great remove in time and space." So much did this become the usual style of thinking about God that the "other" God, the one who spoke to Moses, was actually a source of some discomfort to later theologians. A God of such universal sovereignty has to be almost inconceivably immense, regarded "perhaps in the way today's physicists regard their descriptions of the tiny particles they study, not easily imagined as actual objects, but as models which should not be taken too literally."

Scholars have noted that there are actually two versions of God in the Hebrew scriptures. Each behaves in a somewhat different way, though it is clear both names refer to just one being. Elohim, "God," is universal, faraway, detached, while the other, the "Lord God," Yawheh Elohim, is more down-to-earth and nearby. It is Yahweh who walks in the Garden in the cool of the day and calls out to Adam. He makes clothes for Adam and Eve. There are symptoms of instability in his makeup, though also touches of intimacy. He can equivocate. Elohim, the more transcendent deity, has a chilly side to him, intensifying his remoteness. Dissatisfied with the way things are turning out since the Creation, he tells Noah: "I have determined to put an end to all flesh, for the earth is filled with violence because of them." In less frosty terms, Yahweh is "sorry that he had made humankind and it grieved him to his heart." It was Elohim who said: "I will blot out from the earth the human beings I have created—people together with animals and creeping things and birds of the air, for I am very sorry that I have made them."

Here the mercurial character of the Old Testament God, his alterations of mood and personality, the different names assigned to him, sometimes El, sometimes Yahweh or Elohim, all accentuate the sense of withdrawal, of keeping his distance. He is difficult to read and his choices, the people he prefers over others, can seem arbitrary. In the early nineteenth century scholars found evidence of four different source documents for the first five books of the Old Testament, once thought to have been the work of Moses. The four documents were given alphabetical symbols. The one linked to the divine name Yahweh-Jehovah was called J. That referring to the deity as Elohim, God, was called E. The third and largest, which dealt with priestly matters, was given the letter P, and the writings exclusive to the Book of Deuteronomy were called D. These different sources portray more than one sort of God. Yet although his disposition may vary, a constant is his colossal power, his crushing acts of judgment. The J author attributes to Yahweh terms such as "compassion," "clemency," and "kindness," and shows him backing away from his decision to destroy the life he has made. But in the work of the author known as P, whose God is very much in charge of the universe and radically transcendent, mercy and repentance are conspicuous by their absence.

"Messy monotheism" is how another biblical scholar, Jack Nelson-Pallmeyer, describes these hints of multiple personality. It is confirmation that language lacks the resources to portray a single deity who is perfect, who knows the future, and can do anything he wants. When

Jesus in the New Testament talks about God, which one is he referring to? Is it P's intangible version or is it J's personification? Theology down the ages has got into quite a tangle on this question. Christian monotheism emerged out of a contest over how to explain the discrepancy between a good God and a wicked world. In Old Testament times there were priestly quarrels, some of them quite rowdy, about how to best mediate between human beings who are more or less comprehensible and an ill-defined God for the most part an enigma. There were sharp differences between the descendants of Moses and the descendants of Aaron on this issue. A priest had to perform a tricky balancing act, placating a God extremely jealous of his uniqueness, and at the same time keeping people's respect for his own priestly status. It was of the utmost importance that another disaster like the Babylonian exile be avoided, and that meant faithful adherence to the law.

Curiously, when faced with the riddle of a God of supreme power who oversees a world beset by ungodly calamities, religious thinkers tended to make divine power greater than ever. How to explain, for example, why God let his chosen, his beloved people, the Israelites, go into exile? One answer is to say it was to punish them for their disobedience, as a father might discipline a wayward child "for his own good." Being omnipotent, he could recruit foreign empires to carry out the chastisement on his behalf. The expansion of his power may also account for the fact that God's compassion was militarized within much of the biblical tradition. God shows his care through the violent defeat of his enemies. "The designation of God as Almighty and Omniscient may seem comforting," Nelson-Pallmeyer notes, "but it created enormous difficulties for people of faith trying to make sense of God, life and history." That, of course, was Job's dilemma. He had led a decent life, observed the covenant, kept his side of the bargain. And yet this sovereign deity, wielding immense physical power, was making his life a misery.

Growing more powerful, God did not become any less "other." To some extent Israelite religion, as it developed over the centuries, avoided describing God in human terms. The idea that God actually eats the sacrifices offered to him by his worshippers is phased out: sacrifice is only a "pleasing odor" to him. The eccentric prophet Ezekiel is careful not to repeat other biblical passages which describe God as walking, as he does in the Book of Genesis, or showing human emotions. Aspects of God, such as the divine "face" or the divine "form," are not supposed to be seen. The use of masculine terms and metaphors in talking about Yahweh sat uneasily with the uncreaturely status, though they never quite became obsolete. The older tradition of a God with a human face was established by the writer of Genesis 1:26–27:

> And God said, Let us make man in our image, after our likeness: and let them have dominion over the fish of the sea, and over the fowl of the air, and over the cattle, and over all the earth, and over every creeping thing that creepeth upon the earth.
> So God created man in his own image, in the image of God created he him; male and female created he them.

Even the fiercely iconoclastic Ezekiel, when writing about his vision of God while a captive in Babylon, pictures him, though somewhat gingerly, in human terms: "the likeness of a throne, as the appearance of a sapphire stone: and upon the likeness of the throne was the likeness as the appearance of a man above upon it" (Ez 1:26). Genesis, throwing this convention into reverse, instead of scaling down God to fit our persona, magnifies it so that it can sit on a par with the divine.

The growing asymmetry between the cosmic Hebrew God and mere humans meant he must be approached indirectly, through third parties. There is an odd story in Leviticus (9:1–10) about the inadvisability of freelance attempts at closeness with the Almighty. Two of Aaron's sons, Nadav and Avihu, got carried away at the close of the celebrations for the just-completed Tabernacle or "tent of meeting." The Temple, built after Moses came down from Mount Sinai with the Ten Command-ments, marked a new covenant between the Lord and his chosen people, in spite of their disgraceful flirtation with idolatry. Bursting with euphoric craving for an immediate experience of God, the two sons of Aaron took their censers, put incense in them, and lit them on their own, flouting the rule that only the fire that God sends may be used to make sacrifices. Audaciously, they offered "strange fire" before the Lord in the Tabernacle, a serious breach of protocol, and were promptly struck dead for their pains. It is possible that the pair was a little tipsy at the time, since the Lord later said to Aaron: "do not drink wine nor strong drink, thou, nor thy sons with thee, when ye go into the taberna-cle of the congregation, lest ye die: it shall be a statute forever through-out your generations."

A somewhat similar situation arose during the reign of King Uzziah in Judah in the eighth century BCE. Uzziah was a highly successful statesman who expanded Judah's power and made the country prosper-ous. He was religious, and looked for guidance to a prophet, who urged him to honor and obey God. But his early successes went to his head. He wanted to dispense with the prophetic go-between and become a divine king, with exclusive access to God. One day he went into the Temple in Jerusalem to burn incense on the altar. Some eighty priests blocked his path and told him only priests, the descendants of Aaron, could perform this ceremony. Uzziah already had the censer in his hand and tried to bluster it out with his opponents. At that moment leprosy broke out on his forehead, and he was hustled out of the building. He was leprous until he died, and was never again to enter the Temple.

A class of men and women who did claim an unmediated relation-ship with the Lord were the prophets, and at the time of King Uzziah the greatest evangelistic prophet of them all, Isaiah, was exercising his brilliant gifts as a poet and preacher. Unlike the priests in the sanctuar-ies, whose authority came from having Aaron as an ancestor, the proph-ets could dispense with proof of lineage; they deliberately used language inventively to shock and scandalize. They were improvisers, disturbers of the peace, with messages for the people that were often unsettling and in many cases highly offensive. The prophets, again unlike the priests, could claim a personal audience with God, a proximity unique to themselves, qualifying them for the role of intercessor as well as communicator of divine instructions. A prophet receiving his revelation was totally iso-lated before God. And to validate his exceptional role he needed to tell the story of how he was called to undertake his exacting mission.

In the eighth and seventh centuries BCE, God was apt to address the individual prophet at first meeting in a quite straightforward and per-sonal manner. This involved a complete break with the candidate's past life and leanings. The call was usually prefaced by a vision. But here is an odd fact. Even though the great prophets had one-on-one encounters with the Lord, they give us remarkably little in the way of information as to what sort of being he is. They make terrible, willfully reticent eye-witnesses. The prophet seems maddenly incurious as to what God is "like."

As we have seen, we get little more than vague generalities from Ezekiel, who is so tentative in his portrait of the Almighty as to be quite annoying. Yet Ezekiel comes as close as anyone to describing God. Isaiah had a vision in the Temple in Jerusalem, the same one in which King Uzziah had recently been checked for his presumption, soon after Uzziah died. Like Ezekiel, he sees Yahweh sitting on a throne. Above the throne stand seraphims, each with six wings, two to cover his face, two to cover his feet, and two in order to fly. One seraphim cries out to another: "Holy, holy, holy, is the Lord of Hosts, the whole earth is full of his glory." Isaiah probably came from the ruling class of Judah and is clearly a sophisticated person. He is thunderstruck that he has actually been in the presence of "the King, the Lord of Hosts." But he tells us next to nothing about what he has seen.

"Contrary to popular misconception," according to the Old Testament scholar Gerhard von Rad, "the prophets were not concerned with the being of God, but with the future events which were about to occur in space and time—indeed in Israel's own immediate surroundings. Yet even to the theologian this massive concentration on historical events, as also the complete absence of any sort of 'speculative' inclinations even in those visions where Yahweh is seen in person, must be a source of constant wonder." Yahweh himself discloses very little about his own "being." Karen Armstrong, a best-selling historian of religion, notes that when God appears to Isaiah he is enveloped in an opaque cloud, like the cloud and smoke that hid him from Moses on Mount Sinai, and the cries of "Holy, holy, holy," far from telling us more about him, actually underline the fact of his difference from us. The Hebrew *kaddosh*, translated as "holy," does not mean purity or virtue, but rather "otherness," a radical disconnect. The seraphs, in crying, "God is Other! Other! Other!" were really announcing the fact that a huge metaphysical chasm separates us from the divine. The violent stage effects reinforce the impression of something too devastating to accommodate mortal knowing. Isaiah must have his lips cauterized with fire from the altar before he is fit to make even the brief, awed responses he can muster in reply to the Lord's instructions.

Yahweh, in giving his instructions to Isaiah (6:9–10), speaks with

heavy, all too human sarcasm about the spiritual immaturity of the people to whom the prophet is assigned to preach. "Go, and tell this people,
Hear ye indeed, but understand not; and see ye indeed, but perceive
not," he tells Isaiah. "Make the heart of this people fat, and make their
ears heavy, and shut their eyes, lest they see with their eyes, and hear
with their ears, and understand with their heart, and convert, and be
healed." There is not much mystical illumination here. The message is
that the prophet has a job to do, and no one says it will be easy.

Often we get a feeling of awkwardness in the interviews between
Yahweh and his chosen prophet. It is clear that direct contact can be an
abrasive, jolting experience and highly disconcerting. It is certainly not
something the average person could handle easily, which is why prophets
were needed as a buffer to "bear the brunt of the divine impact." There
is frequently an element of physical violence: howling winds and earthquakes and the loud beating of wings, fire spurting in all directions.
Ezekiel is handed a scroll and told to eat it; it turns to honey in his
mouth. Nonetheless, he was "stunned" for an entire week.

There is irony in the treatment of Yahweh by these exceptional
poetic minds. What meager information they provide about the divinity is given in human metaphors and implies human concerns. To Isaiah, Yahweh is a monarch enthroned. Amos, the shepherd, a nobody
before his calling to prophecy, suggests the Lord has care for the poor
and outcast. "The prophets in an important sense were creating a God
in their own image," says Karen Armstrong. They were projecting
some of their personalities onto an otherness that never quite comes
into focus. The prophet Hosea may have made an unfortunate marriage with a prostitute. Whether or not this was actually the case, his
portrait of God is of someone who entered into "matrimony" with the
people of Israel, who are flagrantly unfaithful to him, go whoring after
other gods, "and love flagons of wine." Hosea complains that the people are no better than prostitutes, consulting idols and unfaithful to the
Lord. The young women turn to adultery and the men consort with
harlots. "All religion must begin with anthropomorphism," Armstrong
comments. "A deity utterly remote from humanity, such as Aristotle's
Unmoved Mover, cannot inspire a spiritual quest. As long as this projection does not become an end in itself, it can be useful and beneficial.
It has to be said that this imaginative portrayal of God in human terms

has inspired a social concern that has not been present in Hinduism. All three of the God-religions have shared the egalitarian and socialist ethic of Amos and Isaiah."

This can lead to complications. Prophets may to some extent depict God in their own image and saddle him with their own concerns, but he need not conform to that portrait. The Book of Job is a fascinating example of God being "other," shrugging off efforts by humans to understand him on their own terms. The Psalms say God puts himself under the law, but in his dealings with poor long-suffering Job he seems to stand outside it. "The Lord may exalt the righteous and humble the wicked," says Jack Miles, "but, then again, he may not. Nothing he says or does need imply more than 'I am who I am.'"

Here we see a developing paradox that will haunt theology for millennia. During the pivotal seventeenth century, when the modern world can be said to have come into being, science describes a physical universe ingeniously constructed to produce a consistent, well-behaved, reasonable system, the "clockwork" cosmos. In such a vision the universe is thoroughly understood and therefore completely predictable. Science is a sort of mediator, putting people in touch with a concept of God as orderly, steadfast, trustworthy, law-abiding. That is not the case, however, if we use the Bible as a literary mediator. We get a God who is sometimes capricious and unreasonable. A curious example is in the Book of Ezekiel (chapter 20), where the Lord refuses to speak to a delegation of the Elders of Israel, even through his prophet. Why? Because the Israelites had practiced "abominations" by worshipping idols of Egypt. When God brought them out of Egypt and put them in the wilderness, "I gave them my statutes, and showed them my judgments, which if a man do, he shall even live in them." But in the wilderness, the house of Israel went rogue, rebelled against his statutes and scorned his judgments. For that reason, the Lord made an odd sort of move. Because the Israelites had disobeyed his rules and ignored his conditions, "Whereupon I gave them also statutes that were not good, and judgments whereby they should not live. And I polluted them in their own gifts, in that they caused to pass through the fire all that openeth the womb, that I might make them desolate, to the end that they might know that I am the Lord."

What is going on here? These last, disturbing verses refer to child sacrifice, the burning of children alive as a conciliatory offering to

Moloch, a Canaanite deity, who insisted on the sacrificial incineration of the firstborn. It is because of this abominable practice that the Lord refuses to speak to the delegation of Elders. But the Lord had seemed to say to Ezekiel that the Israelites sacrificed their offspring in order to please him, to obey his altered rules. This raises the fascinating question, as Jack Miles reads this passage, of whether it is sometimes ethical to disobey God and whether, in doing so, humans are God's ethical superior. "If God is capable of testing mankind by masquerading as a demon," Miles writes, "then paradoxically mankind can only please God and pass the test by defying God."

Read as a piece of literature, the Bible presents Yahweh as a character central to the narrative. But in that case, "the demonic strand in his character, though it does not develop, raises the possibility that the historical sufferings of Israel *are* the crime of God. The writer is willing to imagine, however fleetingly, that God seduced Israel into the very sins that he then punished—all to prove 'that I am Yahweh,' which is to say, to reveal his character, to put himself on display." As we shall see, the mismatch between God's ethics and our own will be fodder for theological speculation down to our modern time. It is a constant roadblock in the way of domesticating the divine.

Relations between the divine and the human in the Old Testament are complex, to say the least. There is a certain disjointedness that seems to be built in at the Creation. Adam and Eve disobey from the start, and within ten generations Yahweh decides that men and women were "evil continually" (see Genesis 6:5). There was hardly a scrap of good in them, and after only 1,656 years they would be wiped out in a flood. Noah and his family are spared, and when a rainbow appears in the sky, God makes a covenant "for perpetual generations" (9:12). But within a handful of generations the descendants of Noah are busy building the Tower of Babel, planned, metaphorically, to reach to the heavens. God then reminds humans of how different he is from them, how futile it is to expect a direct connection, to "speak the same language." He introduces confusion into human tongues and scatters the single human family into different clans, nations, and ethnic types.

Mutiny is a recurring theme in the Bible. During the wilderness years, there are more than a dozen rebellions, a defiant turning to other gods. Moses describes the Israelites, after spending forty years with them, as a "stiffnecked people," who do not deserve the Promised Land. "Remember, and forget not," he tells them, "how thou provokedst the Lord thy God to wrath in the wilderness: from the day that thou didst depart out of the land of Egypt, until ye came unto this place, ye have been rebellious against the Lord." More than once, Yahweh is determined to destroy the "stiffnecked people" and start over with another nation descended from Moses, but each time he changes his mind. He is always giving his insubordinate people another chance so that, to a degree, his compassion and forgiveness compensate for his deepening seclusion.

Some scholars see this withdrawal as inevitable, given the nature of God as well as the character of humanity. If people are told they are made in the image of the divine, they naturally aspire to *be* divine, like King Uzziah. But that is against the rules. And the Old Testament makes it clear that human beings typically are not able to tolerate close proximity to the divine: it is too overwhelming for them. So they resent their inferior status, their failure to be godlike, all the more because it seems to be part of the unalterable order of things. "It is not just that God is good and humans are bad," writes the Hebrew scholar Richard Friedman. "It is rather that God is God and humans are humans. The very nature of the divine is at odds with the very nature of human beings, and perhaps they are incompatible precisely because humans have something of the divine in them." The incompatibility seems to be baked into the cosmic cake, implicit in the very temper of creation.

So the relationship has to be negotiated through third parties, or mediated by some different kind of knowledge, and the rules of communication are drastically altered. Contact is through channels that are neither quite divine nor strictly human. Friedman thinks the implications of God's withdrawal in the Old Testament are just as significant in the New Testament, and in postbiblical Judaism. The arrival of Jesus as mediator, supposedly a monomediator, the One and Only, does not eliminate the need for others, and there is a great deal of confusion as to the role and status of those others.

The Babylonian exile was a turning point in this respect. God had became an international deity, with bigger concerns on his mind than just his chosen people. He was less intimately local. Third parties were needed to act as couriers and envoys. There was a growing interest in intermediary beings. In texts written after the exile, several angels are named, including Michael, Gabriel, and Raphael, who have specific tasks to perform for a busy deity. In contemporary Jewish writings Wisdom emerges as a person, Sophia, rather than as an abstraction. In the Book of Proverbs she speaks of herself as present at the Creation of the world as God's "architect." An influential scholar, Wilhelm Bousset, argued that these trans-humans, while deferential to God, were a hazard to the Jewish principle of one God and one God only; nevertheless, they were part of the context of early Christianity, enabling the fledgling movement to treat Jesus as a celestial being who had existed from all eternity.

The existence of the angelic host was a feature of Jewish literature written after the exile, and may suggest angels were the objects of cult worship. Jewish exorcists drove out demons by speaking the names of angels, which were also included in charms and spells. A very ancient Jewish tradition describes God as the Lord of the Heavenly Host. By the time of the Second Temple, there were numerous writings on angels and demons and in the Torah the Angel of the Lord carries out missions for God to further his plans. Michael was singled out as the guardian angel of the chosen people and represented them in heaven. The archangels were messengers, bringing the word of God to those specially chosen to receive it.

There is little evidence to suggest, however, that monotheism was compromised by this development, or that the good angels took over some of God's power. His omnipotence remained intact. And there was resistance to the idea that they divided God from his creatures. Here is another example of the tendency, even when there is acute anxiety as to whether people really are under divine care and protection, to make God more rather than less powerful. These exalted or semi-divine beings actually increase God's preeminence. God, as sovereign, was often described in terms of the royal courts of the great imperial regimes of the day, where a grand vizier was given authority over the court. The fact that such an official was needed only highlighted the dignity of the emperor and the immensity of his domain. The idea of a principal angel near God's throne, having command over the entire heavenly host, was therefore not so strange.

Neither science nor theology has quite succeeded in banishing angels from the popular imagination, even in our sophisticated age, a state of affairs we shall now examine.

CHAPTER THREE

Angels Come Out of Hiding

Considering therefore the signification of the word
Angel in the Old Testament, and the nature of Dreams
and Visions that happen to men in the ordinary ways
of Nature, I was enclined to this opinion, that Angels
were nothing but supernaturall apparitions raised
by the special and extraordinary operations of God. . . .
But the many places of the New Testament, and our
Saviour's own words, and, in such texts, wherein is
no suspicion of corruption of the Scripture, have
extorted from my feeble Reason, an acknowledgment,
and belief, that there be also Angels substantiall, and
permanent.
    —THOMAS HOBBES

Are we only a parody of the angels? Or were we
created to supplant them?
    —HAROLD BLOOM

IT IS A STRIKING FACT that no considerable Western religion,
however monotheistic, has been able to dispense with the mediating
services of angels. "The spiritual life," says Harold Bloom, a self-
described critic of religion, "whether expressed in worship and prayer,

in private contemplation, or in the arts, needs some kind of vision of the angels."

Today, in the Roman Catholic Church, that need is still being recognized and, within limits, catered to. Cardinal Joseph Ratzinger, then Pope John Paul II's top theologian, now his successor, said a belief in angels is quite acceptable and, indeed, a positive benefit, because it magnifies rather than detracts from our idea of God. "There is a primitive knowledge in man, somehow," he told an interviewer, "that we are not the only spiritual beings. God has filled the world with other spiritual entities who are close to us because his whole world is in the end one single place. They are also an expression of his greatness, of his goodness. In this way angels are part of the Christian conception of the world, part of the breadth of God's creation. They also constitute an immediate and living environment for God, into which we are meant to be drawn."

The tradition of the personal angelic protector, lowest in the accepted ranking of angels, lingers in some quarters of the Roman Church today, whether playfully or in earnest. The writer Flannery O'Connor recalled that in her six years at a Catholic school she developed "anti-angel aggression" when the Sisters told her she had a guardian angel who never left her side. "It was my habit," said O'Connor, "to seclude myself in a locked room every so often and with a fierce (and evil) face whirl around in a circle with my fists knotted, socking the angel." Her dislike of this unwanted assistant was "poisonous." She knew you can't hurt an angel, even by kicking it, "but I would have been happy to know I had dirtied his feathers."

It is not unusual to hear nuns in Catholic schools tell their pupils to "move over in your desk so your guardian angel will have room to sit down." Children are given holy cards with a picture of a guardian angel. In a bookshop in Easton, Maryland, I came across a complete guardian angel kit: a book, printed in China, to bring "hope, comfort and inspiration for everyday miracles"; a Guardian Angel doll, "to remind you that your real angel is close by"; a dozen blessing cards to place in the doll's drawstring pouch for safekeeping; and pictures of angels, modishly dressed, flying up and down a ladder. The book, published by Running Press, Philadelphia and London, was described as "an insightful, full-color, illustrated guide to learning more about your guardian angel's abilities and how you can communicate with your celestial protector." The guardian angel is referred to as either he or she, "although it is perfectly possible for an angel to be androgynous, or perhaps take a form you never expect." A warning is added: "Remember, angels are estetic beings who are not bound by the rules of physical density and mass. So don't be disappointed if your guardian angel doesn't appear to you in traditional angel garb, wings and all."

The Vatican's official view seems to be that belief in angels is well founded, though no one is obliged to hold such a belief. In any case, it lacks the same degree of certainty as the Church's rulings on the role of the Virgin Mary. That certainly adds to the precarious status of these already ambivalent beings, which hover between fact and fiction, between official sanction and wary prevarication. The Catholic doctrine that angels were created by God was a ruling of the Fourth Lateran Council in 1215 CE and the First Vatican Council in 1869–70. According to the Catholic Church calendar, guardian angels have their own feast day on October 2. Before Vatican II, the Latin mass included a prayer that began: "Saint Michael, the Archangel, defend us in battle, be our defense against the wickedness and snares of the devil. May God rebuke him, we humbly pray; and do thou, O prince of the heavenly host, by the power of God, thrust into hell Satan and the other evil spirits who prowl around the world for the ruin of souls. Amen."

In Judaism, Zoroastrianism, and Islam, as well as in Christianity, angels act as mediators between God and humans, carrying revelations and acting as protectors. In Islam, the angels listen to the prayers of ordinary people and offer them to God. There are four archangels in the Islamic hierarchy. Two, Jibril or Gabriel and Kikhail or Michael, are mentioned in the Koran. The others are Azrael, the angel of death, and Israfel, the angel of music, who is multilingual. Jibril revealed the Koran to Mohammad and took the prophet on a tour of paradise. Islam also has Malaika, guardian angels, as well as the Hafazah, specialists who protect against the Jinn and mischievous spirits. A pair of angels go along with each human person throughout life, one on each side. The angel on the left records bad deeds, that on the right, all good intentions and actions. The Huri are female, inhabitants of paradise. The Al-Zabaniya stand guard at the gates of Hell. The angels are not all beatifically good, however. Harut and Marut cannot resist the temptation of sexual adventures. According to an old Islamic tradition, there are seventy angels, select and privileged, who are in charge of the workings of the universe. In Zoroastrianism the supreme God, Ahura Mazda, is attended by six angels, the Amesha Spentas.

In the Hebrew scriptures there is a certain confusion as to the exact identity and metaphysical status of figures who clearly exercise extraordinary spiritual power. They seem to step in and out of divinity without causing undue consternation in those who encounter them. The exact nature of an angel, the question of whether it is humanoid or so deiform as to be virtually indistinguishable from God, is sometimes undecidable. In the Old Testament these winged mediators can be quite ambiguous. Scholars suggest that the original author wrote God himself into the text, and a later editor, amending the story in the aftermath of the Babylonian exile and wanting to protect God's privacy, transformed him into an angel.

In any case, a reader can be excused for wondering whether angels are stand-ins for the Real Thing, or the Real Thing itself, incognito, dressed up in some sort of camouflage. Early on in the Old Testament, when Hagar, Abraham's fugitive mistress, pregnant and on the run from Abraham's wife Sarah, is accosted by a supernatural being at a water hole in the desert, we are told three times that the apparition is the "Angel of the Lord." But this angel speaks with all the authority of the Lord himself, telling Hagar to go back to Sarah, adding: "I will multiply thy seed exceedingly, that it shall not be counted for multitude." Is that God in person? You would think so. But not really, because in the next breath the angel refers to "the Lord" who has heard Hagar's affliction.

Then there is the story of Abraham's grandson Jacob, who, alone on the night before a momentous meeting with his estranged brother Esau, has a wrestling bout that lasts until dawn with a mysterious opponent who declines to give his name. Finally Jacob says to the stranger, who is unable to gain the upper hand over him, "I will not let you go until you bless me." Whereupon the stranger asks his name. "Jacob," is the reply. "Your name will not be called Jacob any more," responds the stranger, "but rather Israel, for you have fought with God and with people and you have prevailed." After being blessed, Jacob concludes that he has seen God face-to-face and names the scene of the struggle *peni-El*, or "face of God." Jacob exults: "I have seen El face to face and I have survived." The prophet Hosea has it both ways, stating that Jacob fought with all his strength against an angel *and* with God.

Abraham's God, El, who may have been a different deity from those of Isaac and Jacob, is sometimes seen in human form, a familiarity later regarded by the Israelites as somehow improper. El seems to be a less frightening presence than Yahweh when Yahweh appeared to Moses on Mount Sinai. He tends to materialize to the accompaniment of extremely dramatic physical effects, including thunder and lightning, high winds and volcanic flare-ups. These serve to keep people at a distance. El is a much quieter, more approachable deity. When he does make an entrance as recognizably human in appearance, the writer is surprisingly offhand in reporting it. In Genesis 18, Abraham notices three strangers walking toward his tent on the plains of Mamre, evidently hot and tired. He invites them to stay for a while to wash their feet and rest, and arranges for a meal to be served under an oak tree. Without any fuss whatever, Abraham has recognized one of the "men" to be the Lord, and it emerges later that the two others were angels, but this seems quite unremarkable and they are not described in any way.

Angels seem to gain, not lose significance as the pre-Christian period enters its later stages. As God becomes a universal cosmic deity, receding into indescribability, the Bible gives the impression that the space he once occupied is empty and needs to be filled. James Kugel sees angels taking up that role, but these angels are not the same as earlier ones. They are more like humans, having a specific name and a definite task to carry out. As such they are increasingly in demand. In the final stages of the Bible narrative, as agents of God, angels are permanent inhabitants of the divine universe. The Dead Sea Scrolls and the books of the Apocrypha, written toward the end of biblical times, show angels as necessities, not luxuries, in the lives of ordinary people. Even relatively trivial matters, not excluding the weather, could be their concern. They have a share in divinity.

"Toward the end of the biblical period and after, God's very remoteness seems to have compromised his standing as the only divine power in the world," Kugel writes. In practice, divinity was once again parceled out to leaders of men. There are stories of such huge Old Testament figures as Enoch, Abraham, Levi, and Moses going up to heaven to see at firsthand the dazzling splendor of God's throne. They seem to be saying that a person must leave the world in order to reach God, a profoundly mystical point of view. God is vast in scale, but he does not take up so much space as to exclude Moses or other biblical heroes from achieving semi-divine status. It may be that Christianity, in its early stages, and especially in its grappling with the labyrinthine concept of the Trinity, was in its own special way trying to "fill the gap left by the God of Old."

Between the Old and the New Testament periods, God had become so irretrievably transcendent that using his name was prohibited and talking about him in anything like human terms was severely frowned upon. There was an intense focus on his supremacy and utter preeminence in the scheme of things. The theologian Paul Tillich called this a watershed chapter in the history of religion, where the need to compensate for the complete unhumanizing of God led to the invention of names that could not be said to have any human connotation; words such as "glory," "height," "heaven." Intended to recover some of his lost concreteness, these terms were for the most part highly abstract. Mediating beings took over the role of the old quasi-divinities. Since the doctrine of monotheism was firmly in place, battle-tested and almost immune to subversion, there was little fear that the one God paradigm could be impaired by the existence of angels and the like. The angels could "come out of hiding," in Tillich's words. Nevertheless, many of them were derived from lower-ranking gods and goddesses of pagan worship tolerated by the early Christians.

There are angels aplenty in the New Testament. Jesus describes himself in heaven as seated on a glorious throne, his holy angels around him. The angel Gabriel appears to Zacharias, a temple priest, telling him that his wife Elizabeth, supposedly barren, will bear him a son, and he is to name the child John. That went against the wishes of the family, that the infant should be named after his father, but Gabriel was insistent on this point. Zacharias, struck dumb by the encounter—not an exceptionally drastic reaction to a visitation from a divine courier—obediently writes "John" on a piece of paper when asked for a suggestion.

Joseph, the Virgin Mary's husband, received a visit from the "Angel of the Lord" not in person but in a dream, and was told that Mary would give birth to a son, who must be named Jesus. In the sixth month of Elizabeth's pregnancy, Gabriel came to tell Mary her son would be the Messiah. When Jesus is born in Bethlehem, the Angel of the Lord shows himself to the shepherds tending their flocks in the fields, surrounded with divine glory and accompanied by "a multitude of the heavenly host." Later, when Jesus goes into the wilderness to be tempted by the devil, he is invited to throw himself down, since "it is written, He shall give his angels charge concerning thee, and in their hands they shall bear thee up, lest at any time thou dash a foot against a stone" (Matthew 4:6). Jesus declines the offer, and when the devil gives up the attempt, "angels came and ministered to him" (4:11).

It is in part the ambivalence of angels—are they more like us than they are like God, or is it the other way round?—that made early Christians cautious. How did they differ from other supernatural beings? There was reason to take a hard look at the whole concept. St. Paul thought laws had been given to the Israelites via the agency of angels, but noted that the Israelites had failed to keep the laws. Thomas Aquinas accepted the existence of angels but was firm in his view that they were in no sense material beings; in his opinion even the basic question of whether angels have bodies was open to debate. And not all angels are virtuous, which makes the situation even more unclear. Some are deceitful and fraudulent, cheating souls and taking them captive, barring the way to the true God. Demons disguise themselves as gods, inducing fantastic hallucinations and apparitions.

Whatever the precise nature of angels, their acceptance in principle by the Roman Catholic Church meant humans were able to venture into a metaphysical space that previously had belonged only to God. The same had been true in the case of the prophets, but prophecy had long been in decline, as we have noted, diminishing in quality after the time of the Babylonian exile in the sixth century BCE, when some of its functions were taken over by wise men and commentators on the Bible. Jesus himself was aware that confidence tricksters were masquerading as latter-day Isaiahs. In Matthew's Gospel (7:15) he is quoted as saying: "Beware of false prophets, which come to you in sheep's clothing, but inwardly they are ravening wolves."

What seems to have happened is that angels, too, have been on a downgrade, forfeiting some of their illustrious "otherness" as they become more homely, more familiar, more involved in the everyday piety of ordinary people. At the time of Newton, the question of whether angels actually had bodies or were purely immaterial and spiritual was not a frivolous debate. The realm of the tangible, of the substantial, was contending for priority over the intangible, thanks in part to the successes of experimental science. "For a brief and exciting period," in the words of the historian Stephen Fallon, "reality was up for grabs." Milton refused to separate corporeal from incorporeal substance. For him body and spirit were not different in kind, so that in his great poem *Paradise Lost* angels eat and drink. They are "tenuously corporeal," similar in substance to a human soul, and evidence that humans could perhaps turn into angels. Milton reduced the distance between the two. "Milton offered corporeal angels to a world in which corporeal existence was increasingly seen not as derivative from the spiritual ground of reality, but as the ground of reality itself," says Fallon. The poet's monistic take on angels "was not a quaint and curious fringe opinion, but a contender for the minds of future Englishmen."

Today, Harold Bloom sees angels as *domesticated*, "divested of their sublimity" by popular culture, and reduced to being easy, and therefore vulgar. The decay set in, Bloom thinks, when the painters of the Italian Renaissance depicted angels as aesthetically alluring humans. There was a certain feminizing of the image of angelic beings.

Angels also declined in importance and spiritual impact as the upper heavens became overpopulated with them. This is a recurring phenomenon. It happened in the case of Gnosticism, a sect that flourished in the latter part of the second century CE and was a great nuisance to the early Christians. Gnosticism's core belief was that reality has a double character. It consists of a spiritual realm, which is good, and a material one, which is bad, and was created bad from the start. Avoiding the conundrum of why a good God would make a depraved world, Gnostics decided that there is no *unmediated* connection between God and the world. God is neither Author nor Supervisor of it. He is utterly distinct from and foreign to the universe, which was the work of inferior powers. A number of Christians, reluctant to make a total break with official doctrine, tried to amend the Gnostic picture by inventing a large portfolio of quasi-divine rulers, or *archons,* who administered the lower parts of "spheres" of the cosmos on their own. The spheres usually number seven, with an extra one situated beyond the cosmos itself, closer to the divine. God is only remotely linked to these lesser beings, who take the blame for the mess the world is in. The wicked *archons*, warders of the cosmic prison, form a barrier between mankind and the uppermost heaven, and try to prevent the souls of humans from escaping.

As critics of bureaucracy are only too well aware, however, once you start creating subordinate officials, they are liable to multiply out of control. Basilides, a second-century Syrian Gnostic teacher, claimed to detect no fewer than 365 intermediary bodies between high heaven and the universe. The more these entities proliferate, the less responsible God is for what happens on planet Earth and, by the same token, the more removed and aloof he tends to be. The Gnostics were great distance-makers, pushing God out, farther and farther, so as to exonerate him from the Huge Mistake that is our world. One can always find a good reason to add one more intermediary to the list.

Similarly with angels. When their numbers undergo an explosion, they are no longer so extraordinary and "other," no longer so exotic. We hear less about the "orders" of angels, a hierarchy or ranking of senior and junior, and more about angels doing social work among the needy. And when angels multiply, there is the potential for discord and rivalry, raising the question, which Gnosticism brought into sharp focus, of whether there is a basic element of the inimical in the universe, of dissonance, a fracture that angels inadequately straddle. The theme of *Something is wrong* enters the cosmic drama, what Franz Kafka's biographer Max Brod describes as "the eternal misunderstanding between God and man." There is a strong sense of God's distance from the world. Mutiny against such a remote authority is a recurring theme in the Bible story. The episode of the Tower of Babel suggests that trying to speak the same "language" as God is a futile enterprise.

Isaac Newton did not deny that angels exist. But they play no explicit part in his theology or his cosmic scheme. In a censorious thrust at Catholic idolatry, he wrote: "Nor may we invoke Angels or the souls of dead men as mediators between God and Man. For as there is one God, there is but one Mediator between God and man, the man Christ Jesus." At the same time, he gave the impression that science, his science, could be a means of encounter with the godhead, closing the gap between human and divine. Frank Manuel, a respected Newton scholar, thinks that Newton, during the three decades of his intense and protracted intellectual activities at Cambridge University, was sustained by a "consciousness of a direct personal relation between himself and God his Father."

Did Newton's science influence his thinking about the character of the Almighty? Or was it the other way round? The debate is still continuing. What is certain is that he believed that a strict theology of one God, all-powerful, with no apparatus of lesser spiritual beings, of saints and angels and goddesses, was beneficial to the scientific enterprise. Newton assigned to God the task of keeping celestial bodies in their proper courses, which in a cosmology current in the Middle Ages had been the responsibility of angels. For Newton the enemy was a metaphysics that cluttered and complexified what should be simple. Angels could be made redundant, but God must not be allowed to disappear— far from it.

The irony, however, was that a number of thinkers in the age following Newton *did* make God disappear. He had been made so simple, so "easy," it was a simple and easy matter to discard him. Newton had given God the job of keeping the universe on an even keel. When it turned out that the universe is self-sufficient, self-correcting, that job was eliminated. There was nothing left for him to do. A refreshing simplicity tended to decline into banality. The suspicion grew that perhaps God was more complex than had been thought, and nature more sophisticated, especially in the simplified popular versions of Newton's system that were circulated to an eager lay audience. Possibly the unity of heaven and earth, of God and man that Newton postulated was too facile, too neat. It occurred to some in the years succeeding the marvelous discoveries of Newton that faith is sometimes at its hardiest, most robust and confident, when the faithful are out of tune with the secular world, and when God's purposes seem at odds with the brute realities of history.

CHAPTER FOUR

The Technical Trinity

As a lifelong critic of poetry, I admire the poem
of the Trinity without loving it.
—HAROLD BLOOM

For NEWTON, the one God of monotheism guaranteed that the world was an intelligible whole. Theology, not so keenly concerned with understanding the universal machine, needed a divinity who revealed himself on a more personal basis. That was quite beyond science's competence to demonstrate. Science could provide evidence of a wise and masterful creator, architect and engineer, but what of his other aspects of love and caring? Theology would need to devise a God still possessing the prized unity, but he would be more complex—indeed, incomprehensibly so in the opinion of some. In the Old Testament this dilemma was not so salient, since God was not only Author of the universe but also an active intervener in the affairs of humans.

Once, at the momentous meeting on Mount Sinai, God spoke his name to Moses. That name was to detach Israel from all other peoples, making it unique. It sealed the covenant by which the Hebrew deity stood ready to help and act, and it gave him a distinct individuality, as a being who makes decisions and plans almost like a human person. The prophets made much of his love and care, his disappointments and grief when let down, at the same time making it clear that he was totally

Other. It was the writers of the priestly tradition in the Old Testament, active during the fifth century BCE, who held a more exalted view of God, deemphasizing his closeness, his warmth and approachability. These authors had a vested interest in creating forms of worship, observances of holy days, linking them to the priesthood. The encounter between God and Moses they saw as a watershed in history: not as a moment of intimacy, but rather as the instant when monotheism became something abstract that could be argued about and "proved."

After the exile, as part of a need to bind together a society shattered by the hardship of living in captivity, strict obedience to the letter of the law was enforced. The law was eternal, existing before the creation of the world, and it applied to every aspect of a person's life. Not a single moment was spent outside the precincts of the law. This totalitarian and rigid system added to the impression that God was reserved and distant, and could not be named. It was the code of the Pharisees, who earned the scorn of Jesus for missing the point of true religion, getting lost in a legal labyrinth where the sun of genuine faith never shone. And it imposed an excruciating burden on the ordinary person. This may be one of the reasons for what Paul Tillich describes as the "controlled despair" seen in the writings of St. Paul, St. Augustine, and Martin Luther.

It was during the seventh century BCE that an assertive monotheism emerged, insisting that Yahweh was the only God. The once-and-for-all statement of this belief was made in the reign of King Josaiah in Judah: "Hear, therefore, Israel: The Lord thy god is one Lord." At the end of the Old Testament period, some of the pagan gods are depicted as demons. After the great age of prophecy ended in about 450 BCE the "spirit," or *ruah* of God, which had referred to the divine inspiration bestowing on prophets the gift of rhetoric, became more significant. Once, spirit had been part of God, his breath which filled creation with teeming life. But in the period 300 to 100 BCE it split off and became a distinct entity, acting as a mediator between God and his human people. *Ruah* was a source of comfort and assurance and was linked to God's Word, the written testimony of his care and goodness in the earlier parts of the Bible. When a version of the Bible was translated into Greek for Greek-speaking Jews in Alexandria, the translators claimed to be writing under the direct inspiration of the spirit. The Word, like the spirit, took on a life of its own.

It is hardly surprising that inklings of the Christian concept of the Trinity have been detected in the evolving roles of these mediators in the Old Testament story. There are multiple aspects of the same God, which take on separate roles and personalities as God recedes. God acts by Word or Wisdom, and by spirit in the Hebrew scriptures. The Old Testament scholar Walter Brueggemann detects in them a "faint insinuation of a Trinitarian dimension." He also draws attention to a disconnect between the wisdom writings, comprising Proverbs, Job, and Ecclesiastes, dated from the fifth century BCE onward, and the rest of the Old Testament. Yahweh seems to have absented himself, leaving Wisdom to be the effective force, in the person of Sophia, Yahweh's firstborn, who acts as God's representative. In the Book of Proverbs 8:22–25, Wisdom declares her ancient origins:

> The Lord possessed me in the beginning of his way, before his works of old. I was set up from everlasting, from the beginning, or ever the earth was. When there were no depths, I was brought forth; when there were no fountains abounding with water. Before the mountains were settled, before the hills was I brought forth.

Wisdom's status is a middling one, straddling the two realms of heaven and earth. In the Book of Job, which postdates the Babylonian exile, Wisdom is an attribute of God who makes her home in the universe. She is elusive. The Bible suggests that only God knows her whereabouts. Like other supernatural beings, she is not identical to God, but not quite distinct from him either. In the Wisdom of Solomon, written near the end of the BCE period, she, unlike the angels, has access to the understanding of God and takes part in the creation of the world. In the Wisdom literature, Sophia seems to be almost an individual person in her own right; but in other episodes she appears to be part of the cosmos she helped to create, an emblem of its design and intelligent structure.

In the New Testament, there is some ambivalence as to whether to equate Jesus with God the Father, and what his own divine status might be. A review of the references to the ways in which Jesus and the Holy Spirit are related shows that Jesus was not referred to as God in the Synoptic Gospels, those of Matthew, Mark, and Luke, which share a common subject matter due in part to an unknown source. The titles given to Jesus—Lord, Saviour, Son of God, Son of Man—provide no final answer. The Fourth Gospel, treated as separate from the other three, does not quote Jesus as saying in so many words that he is God. Raymond Brown, the American biblical scholar, found no evidence that he was called God in the first stages of the New Testament tradition. In the Gospels there are three cases where it is reasonably clear that Jesus is being affirmed as God, and five others where it is only probable, as well as doubtful. In John 20:27–28, Thomas, one of the twelve apostles, needs to touch the wounds of the risen Christ before he will believe. When he does so, he exclaims: "My Lord and my God." But in John 5:19–20, when Jesus is answering an accusation of the Jews—that he claimed God was his Father, making himself equal to God—his words are equivocal: "The Son can do nothing of himself," Jesus replies, "but what he seeth the Father do: for what things soever he doeth, these also doeth the Son likewise. For the Father loveth the Son, and sheweth him all things that himself doeth." The celebrated opening verse of John's Gospel states that the Word was with God and the Word was God, but evades the question of exactly in what way the Word is related to God the Father.

A huge crisis was precipitated in the Christian Church during the fourth century CE, when Arius, a priest in Alexandria, wrote a letter to his bishop making a strict separation between God and Jesus, who he saw as a subordinate, not existing from all eternity, not identical to the Father. The same went for the Holy Spirit, who came into being only when God willed to create him. That suggests God may not be fully known through the other members of the Trinity, because the very idea of creation leaves open the question of whether the Creator is "other" than his creation. The Son may be essentially strange to the Father. The power of God active in the person of Jesus is of a limited, inferior kind. Jesus is in the position of the angels in the Old Testament, who could not look upon the Lord; he cannot see or know the Father fully. Paul Tillich compares this version of the Trinity with the hero cults of the ancient world, where dead men who had performed great deeds were worshipped and called upon for help. It made Jesus a half-God, neither fully God nor fully man: even in heaven he is not quite God. It was this sort of uncertainty as to the true nature of Jesus that accelerated the cult of the Virgin Mary.

What came to be called the Arian controversy led to a general council of more than three hundred bishops in Nicaea, which began in June 325. It was an amazing moment of cooperation between imperial Roman power and the once-ostracized Christians, some of whom came to the meeting with visible mutilations from their years of persecution and torture. The emperor Constantine, gorgeously dressed, threw his weight behind the anti-Arian faction, those standing fast for an orthodox doctrine of the Trinity, hoping that a fully united Christian movement would help bring some cohesion to his brittle empire. He was gracious and genial, sharing a sofa with some of the attending bishops. At one moment he kissed the eyeless cheek of the disfigured Bishop Paphnutius of Egypt, who had been blinded under the emperor Diocletian.

The result of the meeting was a formula, designed to appeal to as many as possible. What was striking about the bishops' decision was its unbiblical flavor, which gave even the anti-Arians pause, borrowing from a service of baptism used in Palestine at the time that christened a child in the name of the Father, Son, and Holy Spirit. The resolution of the Arian controversy was another example of crucial matters of doctrine being derived from liturgy and worship rather than from scripture or theology. It required belief in "One God the Father all powerful, maker of all things seen and unseen, and in one Lord Jesus Christ, the Son of God, the only begotten from the Father, that is from the substance of the Father, God from God, light from light, true God from true God, begotten not made."

This was not the end of the matter. Endless controversy ensued over the exact meaning of "one substance," a term so imprecise as to incite rather than settle the great passions of the clergy. Also, the Nicene formulary, not to be confused with the Nicene Creed, omitted to say that the Holy Ghost was of one substance with the Father. After nearly sixty years of controversy another council, held in Constantinople in 381, decided that the Holy Spirit is to be worshipped with the Father as a single godhead. But the council still avoided explicitly calling the spirit "God" or cosubstantial with the Father. It ruled that Jesus is truly human as well as truly divine. Half a century later, the Council of Ephesus decided that although Jesus is both God and man, he is still one person.

Twenty years after *that*, the Council of Chalcedon, with about five hundred bishops on hand, agreed that Jesus is one person with two natures, divine and human. The idea was that only a being who is fully two-natured would be able to have rapport with both sides and intercede to put sinful men and women right with God. The emperor Marcian circulated legislation making the decision at Chalcedon imperial law and, in a move designed to settle the argument, he prohibited any further public discussion. The embargo on public debate did not please certain factions, especially in Alexandria, a major intellectual center, where the absolute sovereignty of a powerful God was strongly maintained. "Monophysites" was the ungainly term used to refer to those who wanted Jesus to have only one nature. They set up separate Christian communities in Africa, the Middle East, and Persia. Their adherents survive today, as members of the Coptic Orthodox and Ethiopian Orthodox churches.

Defenders of the two natures doctrine said anything less would have been a disaster. If the moderates who wanted to say the Word is "similar," *homoios*, to God rather than *homoousios*, "the same," had won, it would have put God out of reach and entirely detached from humans. Christianity might have become just another religion among many. But an obstinate question arises: If a two-natured Word is the perfect instrument to mediate between us and God, why then does the future story of Christianity contain so many alternative mediators, channels, agents, couriers, third parties, and vehicles linking the realm of the divine with that of our mundane world? Is it that the tremendous efforts at Nicaea and Ephesus and Chalcedon produced a neat and tidy package that satisfied the theologians but proved inadequate for the needs of the common worshipper?

Technical terms were developed in an endeavor to ease the bitter disputes among the parties, but clever wordplay was of no avail in finding a solution to a deep *religious* dilemma. During the medieval period, the sheer complexity of the doctrine of the Trinity led to a decision by theologians that not all members of the Church could be expected to gain explicit knowledge of, and therefore believe in, abstruse formulae whose intricacy was in part a result of making nonintuitive ideas acceptable. This was one reason for the notion of "implicit faith" which Calvin attacked so vehemently, the doctrine that the ordinary person, not being one of the intellectual elite, needs to abandon hope of full knowledge of God and instead blindly rely on the wisdom of the Church authorities.

In fact, the career of the doctrine of the Trinity is one of fascinating and dramatic ups and downs. You could be burned at the stake for disbelieving in it as late as the middle of the sixteenth century in Calvin's Geneva. By the time of Newton, who personally had no use at all for it but kept his opinions to himself, it was little more than a disqualification for public office. Behind this shift in the climate of religious orthodoxy was the fact that the Trinity had lost some of its importance. The Reformation put it on the back burner, highlighting instead the newly contentious issues of God's arbitrary grace and whether the communion bread and wine actually became the body and blood of Christ. But Catherine LaCugna, in a recent well-noted study, argues that as early as the fourth century the Trinity tended to be discussed in isolation from the question of salvation and how it is obtained, and took on an intimidatingly technical and abstract tone quite divorced from the urgent concerns of the ordinary believer. A new Protestant emphasis on the authority of scripture did not increase the appeal of a largely unscriptural doctrine. And human brainpower—the rise of science reinforced this tendency—was replacing the inner light of the Holy Spirit in affirming the literal truth of scripture.

In the eighteenth-century Enlightenment, anti-Trinitarianism became open and more widely propagated. Clever divines, deploying reasoned argument, tried to put it into logical form, but they only succeeded in making it more abstruse. On occasion it actually came in for public ridicule. A late eighteenth-century clergyman gave it as his opinion that the decline of religion was partly due to the "contempt which has for many years been cast on the Holy Spirit." At the time of the scientific revolution several liberal clerics in good standing had flirted with heresy on this matter. They were motivated in part by the vogue for simplicity and plain speaking in religion as in many other kinds of cultural pursuits. Edward Fowler, bishop of Gloucester, who urged preachers to keep it simple, keep it plain, wanted to revise the prayer book so as to make the Athanasian Creed optional. The great "romantic" theologian of the nineteenth century, Friedrich Schleiermacher, relegated the Trinity to a virtual afterthought in his writings. In the next century Karl Barth obstinately held to the position that the sovereignty of God comes first, and only then can we talk about his "three modes of being." Monotheism, Barth said, is God in his sheer uniqueness, his "otherness over against all others, different from all the ridiculous deities that man invents."

Many people today regard the Trinity as a speculation for theological specialists, having little to do with real life. They find the task of believing in a one-dimensional God and living up to that belief day after day daunting enough. The doctrine of the Trinity is scarcely mentioned in modern writings defending Christianity against its critics. One of the more cogent defenses of the doctrine is that, while it is undoubtedly a more complex idea than simple monotheism, it is a means of getting away from a concept of God as first and foremost the maker and owner of the world. Thinking of God as an omnipotent sovereign, we recall, tended to foster the Church's fascination with power at a time when absolutism was in vogue in Europe. It was an invitation for God's creatures to participate in the divine potency, to lord it over mere nature assisted by the discoveries of the new science. The theologian Jürgen Moltmann thinks the Trinity doctrine could be a remedy for the harmful notion that God's relation to the world is a narrow one of power and domination. Moltmann argues that intense monotheism of the kind promoted by Newton, stressing absolute transcendence, and the divine "difference" helped to secularize the world, with the consequence that the human individual, God's image on earth, sets up as a godlike ruler of his environment, a sure prescription for ecological crisis and much else besides.

The image of Jesus, at times highly unstable and apt to alter with fashions in the culture, was a key factor in the changing fortunes of the Trinitarian doctrine. In the fourteenth century, there was a fondness for the warm and tender side of Jesus, the gentle itinerant charmed by childish innocence and playfulness, enjoying the company of quite ordinary and not especially saintly adults. The critic Helen Gardner has described how the sense of the awfulness of the godhead, played up in earlier paintings and sculptures, gave way to "the pathos of the babe born in a stable and the sweetness of the maiden-mother who rocks him in her arms and sings him to sleep." Such soothing images edged out those of the strong and slightly forbidding Son of God, which called attention to the divine nature of Jesus at the expense of his human one. "The mother and her babe, the mother and her dying or dead son,

became supreme embodiments of loving and suffering humanity,"
Gardner claims. Early Christian mosaics, and painters in the Byzantine
tradition, by contrast, portrayed the Virgin as majestic, sovereign, seated
bolt upright on her royal throne, staring straight ahead. Often the infant
Christ in her lap is crowned like a monarch, radiating power rather than
tenderness. In its leaning to sweetness and feminine serenity, the later
art highlighted the rupture between theology and popular devotion in a
most striking way. Gardner calls this breach a "disease" of Christendom
in the later Middle Ages. It is no coincidence that the Renaissance
period was rich in religious art but relatively meager in theology.

One of the side effects of this disconnect between theology and
devotion was a drift into superstition. Alongside the sentimental adora-
tion of a holy babe entirely vulnerable and defenseless came a preoccupation, almost amounting to an obsession, with aids to devotion that
had no basis whatever in scripture, including saints, relics, and pilgrimages, seen as necessary for the faithful to enter heaven. By the middle of
the thirteenth century, the mania for sacred relics was nearly out of control. After the capture of Constantinople, relics, supposedly unique and
precious, swamped Europe. In fact, many were duplicated or triplicated.
There were two heads of John the Baptist in circulation and three
crowns of thorns. One of Jesus' baby teeth, however, remained exclusive. Frederick the Wise, elector of Saxony, Luther's protector and a
model of medieval piety, assembled a fabulous collection of relics, beautifully displayed, in collaboration with Pope Julius II. They included
four intact skeletons from the Society of St. Ursula and the complete
cadaver of a sinless child murdered by King Herod. On a smaller scale
there was a residue of soot from the fiery furnace, a piece of the burning
bush from which God spoke to Moses, a container of the Virgin Mary's
milk, her girdle, and some of the gold and myrrh presented to the baby
Jesus by the Magi.

Philip II of Spain put together an extraordinary collection of 7,422
relics of saints, among them a dozen complete cadavers, 144 heads intact
and 306 legs and arms. Philip personally assisted in the delivery of the
casket of St. Leocadia from the Netherlands to the cathedral in Toledo,
and observed each bone as it was removed and placed on the altar. It was
said that on his deathbed the only way to rouse Philip from his coma was
to call out to an imaginary bystander: "Don't touch the relics!"

Journeys to the shrines of saints in far-flung places were very much the thing to do; sometimes they involved extreme self-abasement. Henry II of Champagne, who borrowed from ten bankers to fit himself out for the Third Crusade, made a pilgrimage to Roc Amadour, stripped down to his shirt, shackled himself in chains, and crawled up 126 steps to the Chapel of Our Lady on his knees. Henry VIII of England, a sophisticate in matters of theology, himself made a pilgrimage to the famous shrine of St. Mary Walsingham, a notorious tourist trap: Erasmus mocked it for its crass commercialism. There were "fashions" in saints, some rising, some fading, and varying according to the revenues taken in at their places of worship.

A cult frenzy of this kind tended to let the teachings of Jesus drift to the sidelines. Many writings of this era show no more than a flimsy connection with the gospel story. There had to be a reaction. Syrupy images and blatant forgeries masquerading as touchstones for the holy were marginalizing the genuine but harsher scenes in scripture. Protestant officialdom needed to do something about this. It met the situation in part by reasserting the magnificence and omnipotence of God and dampening the ardor for a gentler, more sentimental Jesus. Enter Christ the Redeemer, "whose righteousness is the righteousness of his elect," a more intimidating and removed figure, not so in tune with imperfect mortals, being so utterly perfect himself. This Jesus came to rule the consciousness of mainstream Protestantism in the early modern period. Art had a trickier task depicting such a superhuman ideal. As Dr. Johnson put it, in his usual take-it-or-leave-it fashion: art and poetry aim to heighten and intensify what they depict. But "Omnipotence cannot be exalted; Infinity cannot be amplified. Perfection cannot be improved."

There was plenty of material in the Bible to bolster the image of Jesus as power, as someone to fear. One of the chief offenders in this respect is St. Paul, who uses images and metaphors for Jesus it is hard to imagine Jesus himself would have found acceptable. In Romans 2:5, Paul warns that "But after thy hardness and impenitent heart treasurest up unto thyself wrath against the day of wrath and revelation of the righteous judgment of God." More than once, he makes it clear that Jesus will come to take his part in the terrifying day of judgment. In the Second Letter to the Thessalonians 1:8–9, Paul describes how Jesus with his angels will come down from heaven, "In flaming fire taking vengeance on them that know not God, and that obey not the gospel of our Lord Jesus Christ. Who shall be punished with everlasting destruction from the presence of the Lord, and from the glory of his power." In Acts 17:31, he tells the Athenians that God "hath appointed a day, in which he will judge the world in righteousness by that man whom he hath ordained; whereof he hath given assurance unto all men, in that he hath raised him from the dead." There is a reference to "the judgment seat of Christ" in Paul's Second Letter to the Corinthians 5:10.

Jesus states that the actual judging on the day of wrath will be delegated to himself, as if it were too menial a task to be undertaken by God. "For the Father judgeth no man, but hath entrusted all judgment unto the Son," he tells a hostile audience during a feast day in Jerusalem. "And [he] hath given him authority to execute judgment also, because he is the Son of Man"(John 5:22,27). This could be interpreted as meaning that the Son will be *less* merciful than the Father. Again, Jesus warns in John 5:28–29 that "a time is coming when all who are in their graves will hear his voice and come out—those who have done good will rise to live, and those who have done evil will rise to be condemned."

CHAPTER FIVE

The Sweet Deceits of Piety

The whole Frame of the world is the Theatre, and
every creature the stage, the *medium*, the glasse in
which we may see God.
—JOHN DONNE

IT IS THE RISEN JESUS who seems to take on a more forbidding
aspect, more ready to punish without remission, if that is called for, than
the simple author of the Sermon on the Mount. On foot with his disciples he preaches love and forbearance; but in the aftermath of the Crucifixion a darker note comes in, especially in the writings of St. Paul, the
man who undertook the tremendous task of turning a local movement
into a world religion.

Inevitably, as people emphasized Jesus' role as prosecutor of the
unrighteous, they looked for other mediators, less elevated in scale, more
domesticated. Dead saints, whose number was increasing steadily, were
called upon to plead for the souls of the living, as well as figures from
biblical and postbiblical lore. One of these was the archangel Michael.
In addition to his role in the Islamic tradition, Michael was worshipped
as a God by the Chaldeans, and has been identified as the messenger
who stopped Abraham, at the last moment, from sacrificing his son
Isaac. As St. Michael to Europeans, he was depicted as announcing to
the Virgin Mary her imminent death, and an Oriental tale describes

him cutting off the hands of a wicked high priest who tried to tip over Mary's funeral coffin.

In late medieval piety, devotion to the Virgin Mary, the "mediatrix," inspired a profusion of poetry and prose writings celebrating her role as channel to God. Mercy and grace descended on humans, not straight from the top but via her intercession, and through her unique closeness to Christ, prayers were given a hearing. She became almost a fourth member of the Trinity.

A number of passages in the Bible that once had referred to Jesus were extended to Mary. Ambrose Autpert, an eighth-century writer, abbot of the Benedictine Order and an adviser to Charlemagne, called her the ladder of heaven on which God descends to earth, "higher than heaven, deeper than the abyss, one who deserves to be called mistress of the angels, terror of hell, and mother of the nations." In the thirteenth century, the personified figure of Wisdom, who was with God before the creation of the world and who is made to say, "Whoso findeth me findeth life, and shall obtain favor of the Lord," was identified not as Sophia but as Mary.

There came a time when Mary became so transcendent, so powerful, so full of glory—exactly the attributes that had made God seem unapproachable—that supplicants were calling on less exalted members of her family as negotiators for their requests. These included her mother, Anna. As mediatrix between Christ, himself the "unique" mediator, and the faithful, Mary was actually being contacted via yet another go-between, her husband Joseph, who was asked to "render thy spouse, the most blessed Virgin, propitious to us, and obtain from her that we, unworthy though we are, may be adopted as her beloved children."

At the end of the patristic period, which culminated in the age of Pope Gregory the Great and John of Damascus in the eighth century CE, Scholasticism as an attempt to systematize knowledge of the divine was in its infancy and the monastic movement was getting up steam. More people opted to lead a life of pure devotion, and the vogue for Mary and the saints began to intensify and expand. Mary was elevated. Myths and outright frauds were used to advance her reputation. These included the "proto-gospel" of James, circulating in the fourth century. It tells a story of the birth, girlhood, and youth of Mary, setting out to show how she was chosen by God to be the mother of the Messiah. It depicts the birth of Mary as supernatural, like the birth of the prophet Samuel. Mary is confined to her room as a child, so she will not be contaminated by the world. At three years old, she is taken to live in the Temple in Jerusalem. An angel brings her meals every day. Eventually she is given in marriage to an elderly widower, Joseph, who already has grown children of his own. Mark and John, it appears, were the "brothers" of Jesus.

According to the gospel of James, when the infant Jesus is born in Bethlehem, a midwife comes to the grotto where the holy family are quartered. The midwife sees a shining cloud and then a child materializes. Mary is amazed to see that the child was not born in the normal way, even though she herself was taken hurriedly to the grotto suffering from birth pangs. The midwife rushes out to bring a partner named Salome to witness the extraordinary event. In sheer disbelief, Salome insists on a physical inspection to prove it is truly a miracle, saying: "If I do not insert my finger and examine her condition, I will not believe that a virgin has given birth."

In time, not only the monasteries but the ordinary lay person as well assisted in promoting the cult of Mary. The *Te Deum* was adapted to sing the praises of Mary, who sometimes received more adulation than Christ himself. She was in popular demand. During the last phase of the medieval period churches were built in her name, pilgrimages undertaken, festivals organized, masses celebrated, and tales of miracles put about.

By the sixteenth century, Mary as mediatrix was lifted higher than any saint and all the angels. Theology had changed. It was now more abstract, drier, stodgier, tending to put to one side the life and death of Jesus. There emerged a "Protestant scholasticism." J. S. Whale writes of theology in this period as "a millstone about the neck of living religion," with little to choose between the more scriptural Protestant and more philosophical Catholic texts for "sterility and deadness." By contrast, worship took on a more theatrical flavor, with staged spectacles and ritual in which the congregation was treated like an audience at a play. The rosary came into general use and in 1573 Pope Pius V announced the Feast of the Holy Rosary. The Vatican showed little restraint in the creation of new saints, and actively encouraged the worship of Mary.

As a young man, Martin Luther, who was to cleanse the Church of decadent practices and superstitions, had been immersed in the cult of

Mary and the saints. Once, when he accidentally slashed an artery in his leg with his sword while traveling and was at risk of bleeding to death, he called on Mary to heal him. Later on, aware that such forms of mediation had no basis in scripture, Luther recommended a more sober approach, dispensing with some of the "magical" aspects of her career. But he kept a painting of Mary, by an unknown artist, the baby Jesus in her arms, in his study at the Black Cloister. He did, however, balk at portraits of her with her breasts exposed.

What Luther did not support was Mary's tremendously expanded mediating role at the expense of Jesus. He did not like this modish bypassing of Jesus at all. In fact, he thought, far from being a harmless and very natural impulse to seek congenial contact with quasi-human intercessors, it reinforced a tendency to transform Christ the "kindly mediator" into a dreaded and fearful judge who needs to be mollified, unlike the gentler and more domesticated intercessors. The worship of Mary helped, in his view, to marginalize the Trinity. Luther agreed that Mary was indeed assumpted into heaven; but so were other eminences of the faith, including Abraham, Isaac, and Jacob, who would be with God for all eternity. He felt that Christ was the only mediator, domesticated or otherwise.

In 1530, the emperor Charles V assembled princes of the free cities of the Holy Roman Empire in Augsburg to talk about their religious differences. Luther had a statement of faith drafted, known as the Augsburg Confession, which among other things declared that saints may be held in high esteem as a recognition of sturdy faith, but that there was no sanction in the Bible for invoking saints or seeking help from them. "There is one mediator between God and men, Christ Jesus." A Catholic reply, the *Confutation*, couched in a harsh idiom, insisted that the cult of saints as helpers is proper Church doctrine, affirmed by the great St. Augustine himself. It stated that Mary is a "mediator of intercession." In response to *this,* the Lutherans pointed out that Mary never claimed to have the authority to act as a mediator of redemption. That title was given to her later when piety decided to contrast her mildness and maternal femininity with the image of Christ as the dreadful judge. She was seen as more accessible and benevolent; but the Protestants argued that such authority belongs to Christ alone. The Protestant doctrine was that a person is justified by faith and faith alone, not by merit, and that means faith in the one mediator. Mary prays for the Church, that is certain. But does she receive souls in death? Does she win a victory over death? Does she confer life? The Lutherans were acutely aware that if she can do all those things, what is there left for Jesus to do?

In fact, the word "mediator" occurs seldom in Lutheran writings, which is only to be expected if the single mediator is Jesus. What matters is the uniqueness of Christ, his person and his work. No other being fills that space. The gospel mediates, as does the ministry of teaching and sacrament, and it does so steadily and unfailingly. Today, Lutherans do not even regard Christ himself as a go-between in the literal sense, as one who passes on divine favors to mankind or makes appeals to a God who would otherwise be reluctant to comply. The Lutheran Reformation undertook also to modify the meaning of the word "saint," to discourage the idea that a saint is someone who, by living an exemplary life, is enjoying an eternity of bliss, and is open to requests and entreaties. One of the many paradoxes of the Reformation period, when the world of faith seemed turned upside down, was that whereas the need arose for a throng of mediators to give solace to a turbulent, anxious, and unsafe age, Lutherans looked with disapproval on the cult of Mary and the saints in part because they did not provide sufficient spiritual certainty.

Mary's status has had its highs and lows over the centuries, with popular piety and theology pulling in opposite directions. Tertullian, a Roman lawyer before his conversion to Christianity, and author of the notorious statement, "I believe because it is absurd," downplayed Mary's occult powers, perhaps to avoid a comparison with pagan goddesses such as Isis. He blamed women, as incarnations of Eve, for the mess the world was in. But in the fifth century, Cyril, a patriarch of Alexandria, wrote a hymn to Mary as "Mediatrix of the World." The Council of Ephesus was held in a church next to the ancient temple of Diana in 431, at a time when there were still fears of imitating the old goddess cults. The council declared Mary "Godbearer," despite the wishes of the Syrian bishops, who opposed such a title but showed up too late to influence the council members. She was more than human but less than divine, which made her perfect for the function of conduit and intercessor. She acquired cosmic significance, since the birth of Jesus was seen as a cosmic event.

Charlene Spretnak, author of a book on Mary's shifting identity, sees her belated elevation as due to the Church hierarchy catching up with lay sentiment. Crowds stood in the streets next to the church at Ephesus and cheered the decree. "All subsequent elevations of the Virgin follow from this decision," argues Spretnak. "Moreover, all such elevations, even into the mid-twentieth century, followed the same order: the laity routinely preceded the hierarchy in perceiving an expanded sense of Mary's spiritual presence and the increased honor it evokes."

But in the second half of the twentieth century there was a change. The Roman Catholic Church began to reassess Mary's role as bridge between the divine and human. It aimed to take a strictly biblical view, dissipating the mystery surrounding her and deemphasizing some of her unique spiritual gifts. Modern Catholicism still retains the extra-biblical doctrine that Mary was born without Eve's sin and was taken up into heaven body and soul. But the Church, as part of a streamlining of Mary's reputation, took away her mediating powers, granting her status only as intercessor, and thus putting her on the level of a saint. That was accomplished at the Second Vatican Council. Mary Hines, a feminist theologian, brings Mary down to earth by describing her as "our truly human sister," now demoted to regular saint status, with the same warrant as the other saints. Once more, however, there is a dissonance between religious authority and ordinary worshipper. Statues of the Virgin, taken from the altar and placed in an inconspicuous niche during the strong reform of the 1960s, began to make a discreet return, and the theologians did not object.

Catholics are not the only ones ambivalent about Mary's place in heaven. An Easter 2005 issue of *Time* magazine featured her on the cover, over a story reporting that American Protestants are now taking a new look. Women, claimed the accompanying article, turn to Mary as a "strong female role model." Furthermore, there are some 8 million Hispanic immigrants from Catholic countries in America who worship in Protestant congregations, opening up the prospect of a "pro-Marian tipping point" and a reassessment of the actual meaning of the Reformation. One innovation might be to remedy the omission, in Protestant sermons, of a full recognition of Mary's role at the Crucifixion, explicitly noted in John's Gospel, and her last appearance as the only woman identified by name in a largely male gathering in an "upper room."

There had been a popular revival of the Mary cult during the latter part of the nineteenth century in Europe, notably in France. Mary made several personal appearances to French citizens in the decades up to World War I, at sites which became destinations of pilgrimage. The Vatican, up against the new enemy of secular modernism, declined to recognize many of these apparitions as genuine. It nonetheless welcomed the waves of devotion each visitation inspired. Spretnak thinks that the "rather desperate" measure of the doctrine of papal infallibility, introduced in the late nineteenth century and used to glorify the Virgin, was also a device to resist the inroads made by modernism. In 1854, Pope Pius IX defined the immaculate conception of Mary as "revealed dogma necessary to salvation." (Luther had called it only "a pious and pleasing thought.") From then until World War II, millions of petitions poured into the Vatican asking for another infallible papal declaration that Mary had been assumpted into heaven. Pope Pius XII complied in 1950.

At Vatican II, progressive priests and theologians singled out the Marian cult for reassessment. Freudian theory was the vogue in psychology at the time, and it was rather embarrassing, they felt, to be underwriting a mother fixation on a cosmic scale. In addition, during the 1950s the Islamic world started to see itself as a unified, politicized unit, unfriendly to Christianity, which made it all the more urgent for the Catholic Church to fall in line with other religions by dropping some of its "infallible" dogmas. The resulting chapter on Mary in the document *Lumen gentium* that emerged from Vatican II was cautious and circumspect. It noted that traditionally she had been called "Mediatrix," but the council avoided doing so. Pointedly, at the opening of the section on her titles, the document stated: "We have but one Mediator." That phrase was a compromise, but all the more glaring and radical for its hard-fought inclusion. The pope himself had intervened on two occasions, hoping to persuade the progressive elements to retain Mary's title as "Mother of the Church."

Several of the bishops and cardinals watched the closing ceremony of the council "through their tears." It seemed that a one-of-a-kind channel of communication between heaven and earth, open for centuries, was now shut down.

Mary had been a buffer, a fourth force, softening the hard edges of an encounter with divinity. She reduced anxiety, made more prevalent by the emphasis on Christ the magistrate. The historian A. G. Dickens has argued against a popular conception of the Middle Ages as a period when the ordinary lay person, seeking safety for his or her soul in the next life, confidently trusted in the good offices of the Church to make it happen. Dickens argues that such a notion is oversimplified and idealized. Fifteenth-century religion did not display "a childlike gaiety and optimism reminiscent of some sweet group of saints by a Sienese master," he writes. "On the contrary, medieval men and women were faced by quite terrifying views of punishment in the life to come. It was small wonder that they felt more comfortable with the saints than with God, or that they came to regard the Blessed Virgin as a merciful mediatrix forever seeking to placate the Son as judge. The notion of a complete and total removal of sin and a guaranteed salvation were wholly alien to the medieval mind."

Along with Mary, the doctrine of purgatory was a sort of shock absorber for believers fretting about God and the afterlife. The doctrine became official in the middle of the thirteenth century, but its roots go back at least to the early Christians, whose markings on the walls of the caves and catacombs where they made their home are evidence that they said prayers for the souls of the dead to mitigate their punishment. The idea that flawed humans who do not merit eternal damnation must still be purged of their faults in a place of pain is even older, surfacing in the second century BCE in the Book of Maccabees 12:39–46, which Protestants do not recognize as canonical. It states that "it is therefore a holy and wholesome thought to pray for the dead that they may be loosed from sins." Adam's punishment, to work the soil by the sweat of his brow, was likewise intended to cleanse him of his sin. Such a concept was, and still is, congenial to the strong Jewish belief in a just God, witness the ritual prayers for the dead at the Wailing Wall in Jerusalem.

Jacques Le Goff, a historian of the medieval period, has made the intriguing suggestion that purgatory, as a "third zone" between heaven and hell in the topography of the afterworld, reflects the evolving structure of later medieval society. By the year 1200, a middle class had emerged, situated halfway between the nobility and the despised peasant masses, making up at least a quarter of the medieval population. It consisted of burghers in the cities and the gentry and yeomen in the countryside. Enterprising and keen to get on, their position was none-theless much less secure than that of the nobility with their titles and land, or the clergy with their sheltered Church offices. They were anxious between-souls, not blue-blooded enough for the top of the social ladder, but superior to the bulk of the people with no hope of getting a foot on even the lowest rungs. And as in purgatory, saints or angels from the high empyrean would sometimes stroll through, bringing "air or a sweet fragrance" into the noxious fumes of the place, so the middle class could imbibe something of the aristocratic perfume of their social betters. A third force in society, like the third zone in the afterworld, modified relations between the other two.

Late medieval writers like Sir Thomas More did not flinch from portraying the squalor and degradation of purgatory. More dwelt on its brutish tortures and exquisite devices for inflicting maximum pain. Its spiteful, sadistic attendants, by their mere presence, created excruciating discomfort, their company "more horrible than the pain itself." Militant Protestants ridiculed the doctrine of purgatory as being utterly without basis in scripture. They poured scorn on it as an international scam, designed to fill the coffers of the Church by scaring up money, supposedly earning remission of sentence. But in doing so—we should not be surprised by now to hear this—they released a deep strain of popular affection for the doctrine, as being a link with past Christian souls and not merely a monumental swindle.

Not that the rich and powerful took the doctrine of purgatory lightly. On the contrary, they treated it with intense seriousness. Wealthy individuals left money in their wills for multiple masses to be said to reduce the length of the purgatorial sentence. King Henry VII of England specified ten thousand masses and paid above the normal rate to ensure they would be said with proper reverence, and the emperor Charles V left instructions for thirty thousand masses. William Bouwsma has described how priests in the Medici Chapel in Florence collapsed with fatigue saying masses for their deceased rulers. When Edward VI banned liturgical intercession for the dead, the English lost no time in reinstating it once the Catholic Mary came to the throne. A. G. Dickens cites this as an example of a religious practice which may be superstitious and partly corrupt, yet at the same time sincere and edifying.

Queen Elizabeth I had a shrewd sense of popular sentiment—and the middle class in medieval times had been notoriously sentimental. She understood that intellectuals wanted one thing, the ordinary person something quite different. She gave Protestant doctrine to the first and Catholic ceremonial to the second. Elizabeth was a committed Protestant and early on made quite a show of her tastes. On the first Christmas Day of her reign, the presiding priest refused to comply with her orders not to raise the consecrated wafer, whereupon she ostentatiously strode out of her chapel. Her spoken and written statements about her religious preferences are ambiguous, but it is clear she wanted, for personal as well as tactical reasons, some Catholic customs and ceremonial. She preferred the clergy to be celibate and she loved splendid and ornate church music. Nor was she an iconoclast; she issued orders to preserve Church monuments. In a collision with the archbishop of Canterbury and other leading divines, she insisted on retaining a silver cross and two candlesticks "standing altarwise" in the royal chapel.

As the Reformation became more radical, however, it widened the gulf between heaven and earth by abolishing mediators. Gone was the solace of a human voice to hear confession and give absolution. That placed a new strain on believers, unfamiliar and crushing for those not emotionally robust. It is for this reason that one leading scholar has called the Reformation a "failure."

The historian Steven Ozment faults the Reformation for its "naïve" expectation that people can transform themselves morally without external aids, whether by argument or coercion. Such an idea went clean against the teaching of Jesus himself. The Reformers "brought a strange new burden to bear on the consciences of their followers when they instructed them to resolve the awesome problems of sin, death and the devil by simple faith in the Bible and ethical service to their neighbours."

CHAPTER SIX

"Almost Like a Contagious Disease"

Any idea of God, when pursued to its extreme,
becomes insanity. The Idea of a just God absorbs all
justice into it and leaves a depraved Creation.
—STEPHEN MITCHELL

It is remarkable how little Calvin had to say about the
love of God. The divine glory replaces the divine love.
—PAUL TILLICH

Under the strain of Reformation theology, which in time
made even Luther seem reactionary, people became more anxiety-
ridden than ever. The new austere dispensation gave little comfort to
the fretful. In fact, the Protestant success in purging medieval religion
of its nonbiblical elements only brought new and uglier superstitions
to the surface. A revived interest in witchcraft and the occult substi-
tuted for the frowned-upon sacramental magic and pilgrimage piety.
The Reformation, Steven Ozment laments, "foundered on man's
indomitable credulity."

Anxiety and a sense of loneliness are themes that recur again and
again in the story of religion as it emerges from the medieval period into
the modern age. "To the fear of hell inherited from the Christian past,"
wrote Carl Bridenbaugh, "many conjoined a gnawing sense of sin, which

the Reformation, by thrusting heavy responsibility upon the individual, had heightened and exacerbated rather than assuaged. When our modern jeremiahs attribute the uneasiness of the present day to uncertainty about the use of the atom bomb, and the manifold tensions and pressures of a technological existence that induces ulcers, mental illness and even suicide, they might pause to reflect that all of these afflictions taken together do not attain the magnitude of the concern of the seventeeth century man about the state of his soul."

In Christopher Hill's opinion, the Protestant doctrine of the priesthood of all believers, and of predestination to bliss or torment, were symptoms of crisis and struggle, not of normal life in a stable society. Anxiety had lurked behind the late classical interest in angels and invisible protectors. The two Reformation "heavies," Luther and Calvin, have been repeatedly portrayed as "anxious" individuals. Richard Marius, one of Luther's biographers, notes Luther's realization, especially marked in his lectures on the Psalms, that the new Protestant sense of God's opaqueness, his intensified Otherness, could make even the most devout person feel insecure. There is a radical discrepancy between God and humankind. In the case of such incongruity, we are totally in the dark as to his intentions for us and can never hope to enjoy unbroken tranquility. Luther, who faced unflinchingly the paradoxes of his faith, could find no proof that God operates in a consistent way within human society.

Paul Tillich singled out "anxiety" as the primary impulse driving Luther to defy the establishment. It was, he thought, a legacy from the insecurity rampant at the end of the Middle Ages, stemming partly from a sense that one can never amass sufficient good works, make enough sacrifices, to ensure the safety of one's soul. In Tillich's view, this anxiety was "almost like a contagious disease." People went to extreme lengths to obtain God's mercy and rid themselves of bad consciences. Anxiety drove Luther into the cloister, where he learned that no amount of penance, no excesses of asceticism, could guarantee salvation. He anguished over the question: How can I get a merciful God? It was out of that anguish and that question, in Tillich's opinion, that the Reformation began. As a firm advocate of the doctrine of predestination, Luther held that whoever hates sin is already a member of the elect. So if you are tormented by anxiety as to the beastliness of your sins, that is a sign that God has predestined you to eternal bliss. Anxiety is an indicator of election.

As for Calvin, William Bouwsma, author of a justly praised study, calls him "a singularly anxious man, and as a reformer, fearful and troubled." Anxiety impelled Calvin through tremendous exertions and feats of organization. "He was unusually sensitive to anxiety, that of others as well as his own," Bouwsma says. "He brooded over it, and much of what he had to say was consciously intended to soothe a peculiarly anxious generation. Whatever its private sources, nothing bound Calvin more closely to his time than his anxiety." Calvin quaked at the onset of bad weather. In thunderstorms he thought that "God seems to want to shake the world and overturn what was otherwise unassailable, for even the rocks tremble." And when the sea threw up high, frothing waves, it was "so frightening that one's hair stands up on one's head."

The seventeenth-century philosopher Thomas Hobbes, though he had a tough-minded resistance to the superstitions of his time, was notoriously susceptible to disquiet. It wove itself into the fabric of his political thought. He attributed his chronic sense of anxiety to the fact that he was born in 1588, the year of the Spanish Armada, the navy sent by the Catholic Philip of Spain in his bid to remove Elizabeth from the throne of England. This prospect scared his mother so badly that she gave birth to Thomas prematurely. Hobbes described himself and his fear as twins, and his theory of politics made full allowance for the prevalence of fear. It is fear, he wrote, that thrusts people to make a deal by which a sovereign is given power over them. In his *Leviathan*, he wrote that "covenants entred into by fear, in the condition of meer nature, are obligatory."

A Reformed religion abolished many abuses and frauds perpetrated on a credulous society. The question is whether it did very much to eliminate the worries that plagued men and women as to the state of their souls. There is no doubt that the Reformation domesticated powers that had been part of a dubious realm of the occult. But at what price? Steven Ozment sums up the situation:

By comparison with traditional practice, the Reformation radically simplified religious life—a change still strikingly visible today, as a visit to a Catholic mass and a Protestant service will immediately stress. Many people found themselves relieved of much burdensome conventional piety. But the Reformation also posed a new and different spiritual threat for the laity. Although Protestantism had simpler religious rituals, each had suddenly become absolute, its importance enhanced by the reduction of religion to a claimed vital core. This raised the stakes spiritually for devout believers. The slack had gone out of religion, but with it went also, as the passage of time confirmed, some of the familiarity and comfort. The reformers in the end created a version of what they had originally vehemently opposed: an elite religion. But it was elite only spiritually, not in a social sense, and it placed the laity and the clergy on an equal footing in the eyes of God. It was a religion for any and all who would forgo the sweet deceits of traditional piety.

Part of Calvin's insecurity stemmed from his theory of God. In his theology, he enlarged God, making the divine more distant, more powerful, and at the same time more puzzling. God was unfathomable, and that meant the order of nature, too, was in part contingent on the beck of God's inscrutable will. Deep down, Calvin had a strong distrust of the world, which seems connected with a fading sense of the cosmos as finite, something with definite boundaries, a place for everything and everything in its place. Since the end of the thirteenth century, this comforting picture had been dissolving, adding to a growing uneasiness among those trying to make sense of it all. God must surely be rational, but any attempt to understand that rationality encounters only bafflement. Similarly, Luther, while never minimizing the operations of reason in everyday life, was under no illusion that it could produce any certainty about God. The upshot, Richard Marius thinks, was that he "succeeded in making God more mysterious than ever."

Calvin felt strongly that God's most important attribute is his power, though he often writes of God's unending love. He wanted a God who was in full control, just as Hobbes later insisted on a sovereign who commands total obedience. Instead of celebrating the Creation as a work of love pure and simple, Calvin preferred to stress that the word "Creator" means a being of total power and dominion. Paul Tillich agrees: "For Calvin the central doctrine of Christianity is the doctrine of the majesty of God." This attitude helped to give the impression that, at the start of the modern age, the Church found power alluring. A new picture of God appeared among theologians, one that stressed his *potentia absoluta*, his unlimited power and might. Power had always been a divine trait but not the most important one. Now, that aspect came to the fore.

"A strange thing happened about four hundred years ago to Christian thought about God," writes the historian A. J. Conyers. "The way was opened to make power itself the goal or aim of human life." God was redefined in such a fashion as to make him more singular, more simple. The Reformation had taken a good deal of the complexity out of religion, and the tendency was to prefer a dominant one-dimensional divinity as opposed to the three dimensions of the Trinity. "Emphasis gradually but definitely migrated to an idea of God as principally defined by His absolute power," says Conyers. "Earlier the attributes of God would include not only the so-called 'natural' attributes of spirituality, personality, life and infinity (which includes the idea of power), but also the 'moral' attributes of holiness, justice, truthfulness, faithfulness and love."

It is a plausible thesis that the new science, Francis Bacon's idea of knowledge as power, and science's conquest of nature, all helped to prop up this concept of divinity. Historians have called the seventeenth century the Age of Power, crystallized in the absolutism of Louis XIV of France and his first minister, Cardinal Richelieu. Dedicated to creating a balance of power in Europe and resisting the ambitions of the Holy Roman Empire, Richelieu made the state preeminent over the pretensions of the Church. For Conyers, the emergence of secular power brokers like Richelieu marks a turn from a religious vision of society to a secular one, with "power politics edging out the idea of the good society." The principle of toleration, a concept much discussed in reference to post-Reformation Europe, by naturalizing questions of ultimate values in public life helped to concentrate power in the state. This new secular order sharpened rivalry among the various powers, made material prosperity a high priority, and loosened loyalties to "intermediate" institutions such as church and family. Even John Locke, the seventeenth century's leading philosopher of toleration, suggested that tolerance need not affect the absolutist techniques and goals of the central government.

Calvinists, whose fortes were discipline and single-mindedness, applied both to the task of going up against the political setup in France and Britain. They were anti-establishment, but they did not come chiefly from the powerless classes. Militant Protestantism was a lure for aristocrats, wealthy landowners, and merchants, as well as for a number of superior legal brains. Perhaps they had a taste for the absolutist harshness of Calvinism, the stark contrast it made between the sovereignty of God and the relative worthlessness of humankind. For these members of the aristocracy, the conviction that such devoted and idealistic people must be among God's elect provided an extra helping of self-confidence and assertion. Calvinism was an all-consuming way of life. Its adherents harnessed secular power, money, and organizing talent to fight extremely powerful adversaries. The Calvinist Philip II of Spain not only used his formidable resources to crush heretics; a gambler who played for high stakes, he also aspired to expand his power abroad.

The successes of the rebels and the shock of assassinations of Catholic eminences led to a rethinking of the nature of secular power. A prince had to show himself not only untouchable as head of state but also as someone closer to divinity than his subjects. If such elect status could be established, to kill him would be an act of sacrilege. To resolve this dilemma, the theory of the divine right of kings was introduced. This was a radical device whereby God, supreme sovereign, confers absolute power on an earthly ruler, who is then answerable to God alone. The ruler may use his power arbitrarily and unfairly, just as God did with the unfortunate Job. His subjects may only respond like Job: impotently. As the historian Richard Dunn points out, although the theory of divine right may seem bizarre today, it tended to reduce anxiety among the laity at a time when anxiety was rife. For pious Christians needing order and stability amid the turmoil of wars of religion, a doctrine that said kings are God's vice regents could supply some comfort in stressed-out times. It also conveniently made revolution tantamount to a blasphemous assault on God himself. If Catholics embraced it, so too did Protestants, and absolute monarchy became popular in England, in France, and with the Hapsburg princes.

A major question for theology in the Middle Ages was this: Which is more salient in God's nature, his intellect or his will? Which side you came down on in this debate made a difference to your view of the world, and to what extent you thought the world was intelligible. If you thought that God's will, his desire and motive, were the dominant traits, then you were called a "voluntarist." Voluntarists regarded God as a being so absolutely powerful, so free of any curbs or conditions on his will, that even reason itself cannot help you decide what sort of God he might be. The very rules on which reason is based are subject to his will. In the extreme case, even logic goes overboard, because logic imposes constraints on how one can argue, and there are no constraints on God. Omnipotence is omnipotence, said the voluntarists, and our worldly ways of thinking are beside the point. God can will the absurd to happen if he so desires.

If, on the other hand, you decide that God's actions derive from his intellect, that is a different story. It suggests that the universe is rational and can be understood by the human mind. Remember, the Bible says we are made in the image of God. That makes life more predictable, perhaps safer: it is an anxiety reducer. Rational anti-voluntarism certainly sets the stage for the advance of science. Carried to the limit, it could support the view that some things happen by necessity, and that perhaps there is a rational order in the universe that is in no need of God for its existence. Inevitably, that diminishes the sum total of God's power.

A compromise, floated in the medieval period, was to talk about a second kind of divine power. Alongside the *potentia absoluta*, which enabled God to do anything he chooses as long as it does not entail a logical contradiction—he cannot create a triangle with four sides—went the *potentia ordonata*, "ordained power," which is what God can do within the limits he imposed on himself when he created the world, though it does not rule out his personal intervention from time to time. In a famous remark, Albert Einstein once said his task as a physicist was to find out whether God had any choice in his worldmaking; whether, in fact, God's power is not limitless but curtailed to some extent.

Voluntarists tend to recommend that we simply submit to the mystery that is God, without bothering our heads too much about whether the Bible story is plausible or whether we should read it metaphorically or literally. "I believe in order to understand," is the rallying cry of the hard-core voluntarist. There is also an inclination to treat each thing or event as unique and singular, to distrust grand theories and sweeping overviews and all systems with a lifetime warranty. During the Middle

Ages there was a restless impulse to look for orders of nature different from the ones already known so as to explore God's power to the farthest nook and cranny of creation. Divine power, and its inscrutable sovereignty, are at the core of all voluntarist thinking. Luther, in *The Bondage of the Will,* wrote that God "foresees, purposes, and does all things affording to his own immutable, eternal and infallible will. This bombshell knocks 'free will' flat, and utterly shatters it; so that those who want to assert it must either deny my bombshell, or pretend not to notice it, or find some other way of dodging it." Calvin declared of the will, with a pitiless absence of good cheer, that it is "often the messenger of bad news and the herald of terror."

It is demeaning to a voluntarist's conception of God's omnipotence to suggest that he cannot be irrational if he so wills. That is implicit in the version of predestination outlined by Calvin, one of voluntarism's most redoubtable champions. The doctrine was hinted at by St. Paul in several of his writings on salvation and taken up by St. Augustine, whose theme of love was somewhat vitiated by his theory of just wars and his belief that humans are so full of sin they have no right to expect to be saved. The fact that the elect are rescued is a wonderful, unmerited gift of grace; that the rest go to perdition is no more than they deserve. The modern theologian Hans Küng blames St. Augustine for filling the devout with anxiety, generation after generation, and for making a contribution to the fear of demons in the Western Church.

An overemphasis on sheer voluntarism makes the world incalculable. And it tends to increase the distance between God and humankind. St. Thomas Aquinas had taught that God and humans can work together in partnership to win salvation, a collaboration that implied a certain closeness. But the Reformers, especially Luther and Calvin, decided that this infringed on God's majesty and glory, which admits of no rivals or partners. In fact, Calvin seems to have been more and more wedded to the idea of an infallibly powerful God as he grew older. By the middle of the sixteenth century he was giving it greater space in his writings and taking more seriously the belief that the unchosen were predestined to hell, a doctrine called "reprobation." At one point Calvin says, in so many words, that the reason God does not save the damned is that he does not want to, and why he does not want to is his own business.

Calvin disliked the cult of the saints for a similar reason, that it tended to water down God's power, "so that any of their petty saints may claim some part of it for themselves." His theory was that God's enigmatic otherness is an aspect of his potency. For this Calvin has been bracketed with Franz Kafka, the anxious prophet of twentieth-century unease. Kafka had his own disturbing version of predestination, of a willful world irreducible to logic and reason. In his stories authority is powerful, and all the more so for being mysterious and behind-doors. It operates by rules we cannot hope to understand. A character may find himself booked by the police and trapped in the entrails of the law while considering himself a model citizen and having not the faintest idea why he was arrested. For Calvin, all efforts to bridge the gap that separates the divine from the human are of no avail because God's grace is unpredictable. Similarly, Kafka's heroes are in the grip of an unreadable providence, faced with the maddening paradox that "the right perception of any matter and a misunderstanding of the same matter do not wholly exclude each other."

CHAPTER SEVEN

Getting God Wrong

Woe to him who seeks to pour oil upon the waters
when God has brewed them into a gale! Woe to him
who seeks to please rather than to appal! Woe to
him whose good name is more to him than goodness!
Woe to him who, in this world, courts not dishonor!
Woe to him who would not be true, even though to be
false were salvation!

  —HERMAN MELVILLE

I know there is wilde love and joy enough in the world,
as there is wilde thyme and other herbes, but we
would have garden-love and garden-joy, of God's own
planting.

  —THOMAS HOOKER

IN THE SEVENTEENTH century, gateway to the modern world, the science of God and the science of nature moved into as close an embrace, as entangling an alliance, as ever was known. Theology, as we have noted, was often at odds with popular piety. Now it gravitated to the elite and brilliant thinkers, who by means of sophisticated technical devices and dazzling ingenuity presented a new picture of the universe, one calling for rethinking the role of the Creator and even the shape of a reformed society.

This was something new. In the medieval period, theology was an occupation on its own. It led a sheltered existence, distinct from other kinds of knowledge, and refused to consort with seekers of wisdom in such disciplines as astronomy or mathematics. For a time, science was a handmaid to theology, a downstairs to its upstairs, until the twelfth century, when troubling doubts arose as to whether science and religion could be of one voice when it came to understanding the cosmos and its creator. Aristotle's theory of the universe as eternal, without beginning or end, together with a God who was only a First Cause and knew nothing of the world, came under increasing suspicion.

Theology's privileged status began to weaken in the age of the first Queen Elizabeth of England, when, at the universities, studies that had been taught by men in holy orders were thrown open to secular tutors. While educated lay people more and more often crossed the borderline between the sacred and the profane, theology lost its artificially created security and people were urged to read the Bible on their own. A core belief of the Reformers was that Adam's sin had crippled the human intellect, especially its understanding of God. But that did not inhibit science from making headway under its own steam, without the imprimatur of the Church. Even Luther and Calvin did not require investigators to be in possession of God's special grace. They might be destined for hell eventually, but meanwhile they could do some useful scientific work.

That man could explore the universe without the guidance of the clergy helped to promote the autonomy of science and cut the umbilical cord with theology. The break between God science and the natural sciences was not violent by any means. In fact, it took place with the blessing of the theologians. Roman Catholics, whose influence waned in England after Queen Mary died, had a more optimistic view of the effects of the Fall on the human ability to think straight. According to the official Catholic homilies, memory, intelligence, and other worldly faculties had only been "maimed" by the unfortunate incident in the Garden of Eden. Such confidence in the human power of reason, however stilted, made it possible to pursue the scientific adventure with a fair amount of optimism.

Against this setting, Isaac Newton's discoveries exploded with maximum percussion. Newton was profoundly religious, though not a conscientious churchgoer, and his particular, very personal slant on the Protestant faith cannot be disconnected from his scientific ideas. In fact, it may not be an exaggeration to say that Newton was one of the first to look at science as a *mediator* between God and the world. His notebooks hint at a temporary breakdown in cordial relations with the deity in his youth, but there is no doubt that, in revealing the hidden engineering of the cosmos, with its astonishing harmonies and elegant design, Newton felt he had come closer to the presence of God. Declaring oneself nearer to God in an era dominated by the Church was risky business, of course. Encouraged by Newton, rebel spirits who chafed at the notion that the priestly hierarchy held the keys to interpretation of the divine began to see ways of justifying their own interpretations. In time, Newtonian theology would be caricatured and distorted by well-meaning admirers.

Newton was very much a voluntarist, contradictory as that may seem at first glance. Newton's science, as opposed to his speculations on various nonscientific matters, describes a universe that is wholly regular and predictable. It is a clockwork system, operating according to fixed rules and describable in terms of bits of matter in motion. What room is there for a willful God when the world is a machine running by its own laws? Yet room was made. If matter is passive, inert, unable to move by itself, laws are not enough. There must be impetus. Particles of matter are not propelled by some mysterious inner *animus* or spirit, or by phantom powers. God makes them move, as befits his role as supreme governor of the universe. He is in full control, active at all times and in all places, sustaining the fragments of stuff that make up the world of physical nature. Such a view could thus, in the words of the historian of science John Hedley Brooke, "purge the world of spirits, demons, angels, vital principles and all the other cosmic junk which was commonly thought to mediate between the forces of heaven and earth." It was a hugely simplifying move.

Newton was the flagship genius of an age of genius, a scientist of enormous mental stretch, a blend of daring flights of imagination and a cautious sense of physical realities on earth. Newton's unique greatness, his complete ascendancy as an extraordinary mind, was recognized during his lifetime, even by those who had felt the lash of his jealousy, the sting of his occasional high-handedness. But he was not simply a towering scientific force. Newton spent immense amounts of time studying ancient history and biblical chronology, developing distinct and quite unorthodox theories of the divine attributes. He read voraciously in theology, making prolific notes into his old age. In good Protestant fashion he tried to get behind official interpretations of the Trinity, and commented extensively on such puzzling texts as the Book of Daniel and the Book of Revelation, proving himself an authority on scripture. Newton wanted resolution of the big question: What sort of being is God? He made it clear that his scientific work was intended to assist in this enterprise.

It was of the utmost importance for Newton that God be sovereign, supreme and uniquely powerful. In his mind, both science and Protestantism demanded it. Here the figure of Calvin looms once again. There is no doubt that Newton read this austere and forbidding Reformation leader. Only eight books from Newton's library which bear his signature and the inscription "Trinity College Cambridge," are still extant. One of them is Calvin's *Institutes of the Christian Religion,* published in Geneva in 1561. It is reasonable to assume that it left some mark on Newton's thinking. Richard Westfall, author of one of the best biographies of the great scientist, considers that for Newton, as for Calvin, "the dominion of God was primary, more important than compassion or love." That is why Newton spent so much time and effort investigating how close was the fit between prophecy and subsequent events. The better the fit, the clearer it was that God exercises dominion over history as well as over nature.

How strange, then, that this monumental figure, head and shoulders above his contemporaries, profoundly religious, tireless in stripping away the sophistries that clogged insight into the true faith, should today be singled out as emblematic of an age that "got God wrong." Yet such is the case. It was for his theology, not for his science, that one of the most respected of modern Newton scholars called him "one of history's great losers." Another historian, William Placher, who considers today's images of God simplistic, looks to Newton's century as a time when the rot set in. Part of the trouble was the language used to describe God. Placher feels that at a time when theologians were busy emphasizing God's omnipotence and distance, the language used to describe him was entirely too much like the language used to describe *us*. Newton's adjectives, applied to a deity of awesome grandeur and power, were no different from those used to characterize human eminences. Divine power is not so different from the power of princes. That recalled the wholly amoral power philosophy of Machiavelli. Newton's great rival, Leibniz, who stressed divine intellect over divine will, thought that God acts according to reasons our minds can grasp: he seemed to think that God's motives are our motives. The fallacy of that conjecture was a lesson poor Job had to learn the uncomfortable way.

According to Placher, the Western world entered a "dramatically different era" when it came to ideas of God in the age of Newton. One reason is that seventeenth-century thinkers—with some justification—began to be more optimistic about human brainpower than Calvin ever was. It was not just that science was making spectacular strides. Calvin conceded that scientists did not need to be a spiritual elite. There was also increased confidence that the human mind could understand God and what he was doing in the world. At the same time, there were tighter standards of proof and therefore stricter rules as to what is reasonable to say about religion. Such constraints, Placher thinks, "led theology astray." The resort to reason, though reinforced by a certain buoyancy of the scientific mentality, may have owed something to a mood of desperation in Europe at the time, due to the stress of coping with thirty years of war with its appalling loss of life, a civil war in England, an economic slump, and vicious eruptions of the plague. Amid so much uncertainty it could be a relief to discuss rationally the great enigma of what and why God is, and therefore to reduce him to a human mental scale. Even calling God "transcendent," meaning he is removed and unaffected by the world, violates the older rule that his properties cannot be defined by *any* terms that we can understand, and, in this case, by a term that is disliked and resisted by many of today's religious thinkers.

A culture dazzled by a new science that takes the complex and turns it into something relatively simple is inclined to prefer a language that is relatively simple. The seventeenth century liked "clear and distinct" ideas. This new, enlightened society had said farewell to the medieval cosmos with its symbolic codes put in place by God, and its multiple meanings. A taste for plain talk in religion tended to make God, his grace and his revelation, less foreign and less strange. In thinking about God, many Christian writers amended the medieval concept of his "otherness" to one of his actual distance from the world. They set his "out there" against the possibility that he is "in here." The more he is out, the less he is in, and vice versa. This suggests God is to be found in a certain place, either near or far off, and not, as the early Reformers held, completely beyond all attempts to locate him or to say how he stands *vis-à-vis* the world. According to them, we can't say, "Where is he?" any more than we can say, "What is he?" But thinkers in the age of Newton did ask such questions. They wanted to know where they could put God in the universe he created, and what particular things he personally managed, as opposed to the unattended ticking of the clockwork that runs in obedience to immutable laws.

As space, under Newton's scheme, became absolute, smooth and continuous, empty of symbols, extending endlessly along all points of the compass, it offered no obvious site for God to occupy. He was either "nowhere or everywhere." But the inclination was to give his relation to the world a definite, physical location. He was partway to becoming embodied in a universe from which science was squeezing out all mystery. It is Placher's thesis that once God became "transparent," describable in clear and distinct terms, identified as this rather than that, perhaps even assigned specific tasks and acquiring features, in a word, *domesticated*, it was no difficult feat to do away with him altogether. "A God with a definable location," Placher sums up, "would be likely to be a God who did *some* things rather than being the principle behind *all* things. The more things in the world that science could explain in other terms, therefore, the less potential need there would be for such a God."

A God who does not do everything is a candidate for doing special things: in other words, miracles. The question of whether he steps in to perform supernatural wonder works was one earlier theologians would not have dreamed of asking. Their point of view has been beautifully expressed in modern times by E. B. White in his classic *Charlotte's Web*. This is the story of a magic spider, Charlotte, who writes love notes in her web in an attempt to save her friend Wilbur the pig from the knife. The little girl, Fern, is the first to notice the words "Some Pig" inscribed on Charlotte's web, a slogan which exerts an otherworldly fascination as it "glistened in the light and made a pattern of loveliness and mystery, like a delicate veil."

"We have received a sign," exclaims Fern's uncle Homer on sighting the strange message. "A miracle has happened on this farm." Fern's mother, Mrs. Arable, however, is disconcerted to see her daughter in such an excited state. She consults a physician, the profusely bearded Dr. Dorian. "Do you understand how there could be any writing in a spider's web?" is her opening gambit in the doctor's office. "Oh, no," Dorian replies. "I don't understand it. But for that matter I don't understand how a spider learned to spin a web in the first place. When the words appeared, everyone said they were a miracle. But nobody pointed out that the web itself is a miracle." Mrs. Arable asserts that a web is just a web, nothing special about it, but the doctor stops her dead. "Ever tried to spin one?" he enquires. She shifts uneasily in her chair. "No, but I can crochet a doily and I can knit a sock." Dorian promptly crushes her by pointing out that someone taught her to do both, whereas nobody taught a spider to spin a web. That makes it a miracle. Mrs. Arable is compelled, reluctantly, to agree, adding, however, that she does not understand it and does not like what she does not understand. "I'm a doctor," responds Dorian. "Doctors are supposed to understand everything. But I don't understand everything, and I don't intend to let it worry me."

Medieval thinkers were Dr. Dorians. They were content to live with the mystery of a universe in which God's activity permeated all that is; it made little sense to ask if he is responsible for this but not for that. It was different in the climate of modernity, when the new science reinforced a taste for plain speech and words that did not equivocate. Writers on religion tended to insist that God does intervene on occasion to infringe or override the laws of that very science, but rarely; the rest of the time the universe obeys the laws and is nonmiraculous. It was but a short step down the slippery slope of skepticism to isolate the "miracles," weigh their credibility, ruthlessly examine the evidence, probe witnesses, and then dismiss them as not up to the standards of accuracy set by the stellar brains of the time.

There was little consensus on just where to put God in the seventeenth century, but many thought he ought to be put somewhere. It was a matter of pinning him down, which earlier theologians had considered a dangerous move. God might be immaterial but he had, in the jargon of the time, "extension." He took up space. It was part of his perfection and independence. Only if a spiritual power is extended, if it has dimension, can there be actual forces operating in the universe. This was teetering on the brink of saying God has a body. It shows, according to the Stanford historian Amos Funkenstein, that "the absolute medieval commitment to an idea of God radically purged of all material connotations, however abstract and remote, was broken in the seventeenth century. Medieval theology, in most of its varieties, viewed with intense suspicion any doctrine that took God's presence in the world too literally. So much was this the case that not only physical predicates, but also abstract ones such as goodness, truth, power, and even existence were at times considered an illicit mode of speech." The new transparency of God, the idea, more sophisticated than it sounds, of the "body of God," made such squeamishness obsolete. God might remain to many largely unreadable, but what can be known about him is at least clear, definite, unequivocal, and for that reason, more familiar because more describable in human terms. The worker of occasional miracles, paradoxically, is more domesticated than the mysterious being whose creation is everywhere a miracle. Putting him in a definite location is a first step to developing a definitive portrait.

Newton himself did not decide once and for all where to put God. He attempted with his scientific undertakings to introduce a God of such overwhelming force that it became difficult to exist in a universe under such dominant sovereignty. Newton was so eager to ensure that God still had a function in a universe that obeyed the laws of physics that he ascribed many other natural phenomena to God. By setting a scientific horizon beyond which God was the ruler, Newton almost guaranteed that intrepid scientific explorers would push that horizon to the point where it seemed God had no place or function in the universe at all.

Just why Newton placed such a tremendous emphasis on the *power* of God is not exactly clear. There is no shortage of conjectures. We might mention one curious and rather exotic influence that is a relatively recent addition to Newton lore: the star Jewish intellectual of the Middle Ages, Moses Maimonides. Born in Córdoba, in southern Spain, in 1135, Maimonides was an uncompromising monotheist and an intense enemy of idolatry, the cardinal sin, which included not only all representations of God in paint and sculpture but also human ideas about God that get him wrong. If we have an incorrect concept of him, we have made him into an idol.

In 1936, Abraham Yahuda, a wealthy Palestinian Jew, a friend of Einstein's and a respected scholar of Hebrew medieval writings, bought a slew of Newton's theological manuscripts at Sotheby's auction house in London. They amounted to more than 2 million words in all. The bundle contained several works by Maimonides and a short essay by Newton in Latin entitled "On Maimonides." Lord Keynes, the economist, was also at the sale and purchased some unsorted Newton papers, mainly on alchemy. Yahuda remarked to Keynes that Newton was "probably" a follower of Maimonides. Keynes was of a like mind, saying that a monotheism as radical as Newton's went well beyond mere anti-Trinitarianism. It hinted, he thought, at a Judaic connection, an Old Testament bias.

Richard Popkin, a Newton scholar at the University of California who has studied the Yahuda manuscripts at Cambridge University, considers that Newton's stress on God's dominion, omnipresence, and transcendence "has a Judaic ring to it." Popkin notes that Maimonides became popular in the early modern period as part of a rediscovery of Judaica, due to its use in debates between Catholics and Protestants and also because Protestants, unlike Catholics, who could boast Thomas Aquinas as a superstar thinker of their own, lacked an intellectual of equal caliber able to wed science to philosophy and religion for the seventeenth century. Maimonides became the Protestants' Aquinas.

For Maimonides, monotheism was not just a doctrine but a truth so precious a person could undergo martyrdom for it. God is single and simple, a perfect unity. He is unique, there is no one and nothing like him, nor will there ever be. He does not fit any of the mental categories we use to describe other people. Any attempt to conceive him *must* keep intact his inimitable, utter uniqueness. There cannot be a hint of likeness to even the most admirable and upright individual. A possible device for accomplishing this difficult feat would be, not to speak in terms of what God is, but rather in terms of what he is not. Not "God is powerful," but "God is not weak." In that way we avoid attributing a human quality to him. Instead of "God is wise," we might say, "God is not stupid." For "God is good," "God is not wicked." By not attaching, but rather taking away, such characteristics from the deity, his radical otherness is left undisturbed.

"Whether we are talking about wisdom, power, goodness or anything else," the scholar of Judaism Kenneth Seeskin writes, "God is beyond the categories we use to describe things in the physical world. He is not a bigger, stronger, better or more intelligent person than we are. That much could have been said about Athena. Rather, he is a being of a completely different order." How, then, can we worship or pray to a being who beggars description in this way? Maimonides replies that even if we are unable to know God's essence, we can gain some understanding of him through what he does, what effects, familiar to us as humans in the world, issue from his totally unfamiliar being. He is single, but his effects, his works, are multiple. Every attribute that is found in the "books of the deity," Maimonides said, "is therefore an attribute of His action and not an attribute of His essence." "It is as if God is a black box," Seeskin writes. "We have some acquaintance with the things that come out of the box, but not with the mechanism by which they are generated."

Newton's monotheism needs no emphasizing. In his *Principia,* he wrote a frenzied endorsement of God's omnipotence in which the word "dominion" or its equivalent occurs six times in four sentences. Elsewhere he states that it is not being of one substance with the Father that merits our worship, but "power and dominion." For that reason, there are to be no go-betweens arbitrating God's power. That is idolatry, papistry. Newton did not go all the way with Maimonides in renouncing all labels and metaphors as descriptions of his God. He used words like "superintendent," "monarch," "pantocrator" to express his fixation on God's power. He did not hesitate to call God "wise," and he certainly did not refrain from describing him as powerful. But in one respect he did reflect the Jewish thinker's views. Newton's insistence that God is active in the universe was part and parcel of his belief that we can know him by what he does rather than by guessing at what he is. That means we must give God something to do, must keep open apertures through which he can act on the operations of the physical world. And that, for a scientist as well as for a theologian, is risky business. We shall see that some of today's brilliant physicists have taken this same hazardous road.

In Newton's conception of the universe we can never be certain, at any moment, whether God is working his will via the normal operations of nature or using his miraculous powers to intervene in an extraordinary manner. Do we *want* to be certain? The scientist and natural theologian Samuel Clarke, who spoke for Newton, said the world "depends every moment on some superior Being for the preservation of its frame; and all the great motions in it are caused by some immaterial power." Leibniz, in a very bad-tempered series of letters, disputed this view that God needs to tinker with his cosmos instead of letting it run on its own. But it is clear that the Newton faction and the Leibniz faction were not just arguing about a point of astronomical physics. Each was talking about a different kind of God. And that is partly why seventeenth-century *science* threw theology into confusion. The theologians had been handed scientific reasons to affirm the existence of a God, but there arose a dickens of a muddle as to what he does, where he acts, and to what extent he has been domesticated.

If all mediators have been banished, it is the one God who acts, alone, in the affairs of the world. Newton, late in his career, took a more austere stance on the supreme majesty of God the Father and the dependent status of Jesus. He minimized the role of Christ in his writings. Jesus was a prophet, for sure, along the lines of Moses, sent to reinstate the original, pure worship of God, which had been tarnished by the veneration of false idols. The doctrine of the Trinity, lynchpin of John Donne's theology, was a kind of superstition, a form of idolatry. The Trinity was a mystery, and Newton did not like mysteries for their own sake. If we give to Jesus all the worship due to the Father we are at risk of making two creators and of being guilty of polytheism. "We must believe that there is *one God* or supreme Monarch that we may fear and obey him and keep his laws and give him honor and glory," was Newton's stern instruction. "We must believe that he is *pantocrator*, Lord of all things with an irresistible and boundless power and dominion that we may not hope to escape if we rebel and set up other gods or transgress the laws of his monarchy."

Newton's stern Commander, owner of the universe, must be praised for his effects and understood through what he does in the universe. If we stick to that simple point, all the metaphysical clutter that led Christianity astray and promoted a gross idolatry disappears. It is crucial, therefore, that there *is* something for God to do. That was Newton's firm belief. It was, ironically, a factor in the failure of his great ambition to replenish an exhausted religion by leading it back to its pristine origins. Instead, Newtonian thought contributed to the rising atheism of the eighteenth century.

CHAPTER EIGHT

Strangeness and the Quest for the Divine

George Bernard Shaw defines Don Juan shrewdly as
that always popular and fascinating figure: the enemy
of God. Prometheus was such, and even the God-
obsessed Milton made a Satan who escaped all bounds
and became protagonist in what had been predicated
as God's show. And, of course, enemies of God are not
the product of Godlessness but arise out of the deepest
of religious motives: protest against the inadequacy
and limitations of prevailing conceptions of God.
—EDMUND FULLER

DESPITE NEWTON'S BEST INTENTIONS, the new scientific and
rational universe curtailed rather than expanded God's reach throughout the world. It made sense for physicist-theologians, keen to find a
place for an interventionist deity who does certain things but not all
things, to look at regions of uncertainty, of strangeness, in the structure
of the material world. Where strangeness is, Tertullian had suggested
centuries earlier, that is the place to expect an encounter with the
divine. Thinkers both ancient and especially modern have learned to
beware of a "God of the Gaps," using divinity to fill in the missing bits
of our knowledge of how the world works. Today, it is argued that lacu-
nas of ambiguity where the normal rules of physics break down are not

"gaps" in the old sense, not temporary apertures of ignorance to be closed eventually by the arrival of new knowledge, but genuine, permanent openings through which the divine might make an entrance. Here again we find scientists reinventing God to fit their latest theories of nature. In this case, science purports to have found a subtle, discreet, and self-limiting deity hiding in the uncertainties where nature does not conform to the established methods of discovery.

Today, possible points of entry are in the indeterminacies of the quantum world and in the open complexity of chaos theory, where very simple equations can produce highly complex and unpredictable results. Where there are loopholes created by an absence *in principle* of complete knowledge, God may squeeze his way in, or so say writers intent on justifying the "God option" in modern physics. There is a voluminous literature on this topic, some of it violently at odds with common sense. More on that later.

In Newton's system, uncertainty is present in part because of his overwhelming voluntarism, linked to his belief in the total omnipotence of the Creator to act as he wills. In fact, the idea of God's omnipotence helped to support a physical worldview that contained a fair amount of "looseness," of cloudiness and ambiguity. It avoided tying God down to a single way of doing things, and left voids into which he could enter. Remember, seventy years before Newton, Galileo had come up against this theological preference for a blurry science of nature in a dramatic way. Galileo paid what may have been lip service to the doctrine of omnipotence. One of the characters in his *Dialogue Concerning the Two Chief World Systems* who seems to speak his own mind remarks that the existence of creatures on the Moon is required by the "richness of nature and the omnipotence of its Creator and Governor." But this did not appear to go very deep, and his resistance to the official interpretation of that doctrine was to be his downfall.

The pivotal crisis that got Galileo into hot water with the Church, however, was not exclusively his mathematical "proof" that the Earth moves round the Sun. There was a popular sentiment at the time that mathematics was a kind of sorcery, a bag of tricks that could be used to prove anything and everything. Such had been hinted at in a preface to Copernicus's *De revolutionibus*, on the motion of the planets. In 1614 a Dominican priest, Tommaso Caccini, preached a sermon in the

Church of Santa Maria Novella in Florence in which he stated that mathematics was a diabolical art and mathematicians, as disseminators of heresies, should be ostracized from all decent society. Galileo, for his part, asserted that the book of nature was written in the language of mathematics.

Cardinal Bellarmine, chief theologian at the Vatican, had ruled that there is no harm in surmising that the Earth is in motion, as long as it is no more than a mathematical device to make a good fit with observations. It is quite another matter to insist, without definite proof, that this implausible conjecture is a physical fact, unalterable, completely separate from the wizardry of numbers and geometry. That would be to injure faith. In Bellarmine's opinion, Galileo had produced no such proof.

To high-handedly exclude any conceivable alternative theory—something Galileo was temperamentally inclined to do—would be, in the view of Pope Urban VIII, to limit God to just one way of organizing the heavens, a patent affront to the doctrine of divine omnipotence. That was not all. Galileo had also given the impression that nature has a special kind of power in her own right, which implied the heretical idea that she is in competition with God's might and glory. In a Letter to the Grand Duchess Christina of Lorraine, Galileo made it clear that, for him, nature is somewhat enigmatic and aloof, and does not go out of her way to reveal her operations to human curiosity. It was a counterpart to the remote and hidden deity of the Old Testament. Scripture, Galileo said, is often allegorical and needs interpreters to tease out the true meaning. Nature, on the other hand, is her own interpreter, and never makes a mistake. Remember, Galileo's actual words were that nature is "inexorable and immutable; she never transgresses the laws imposed on her, or cares a whit whether her abstruse reasons and methods of operation are understandable to men."

With this statement, Galileo was going against well-established Church doctrine. Contested passages in the Bible may contain more than one "truth" and commentators must try to extract them. The ordinary reader needs the help of experts to mediate the intended meaning; there is no relying on the literal sense of the text. Divine communication is indirect, filtered through minds qualified to understand it. But Galileo was saying that in the world of physical nature there is one truth and one only: the "literal" truth. No mediators or interpreters are required to suggest other meanings, or even keep open possible options. Before he became Pope Urban VIII, Maffeo Barbarini, then a cardinal with the Holy Office, made it clear he was unhappy with the theories of the fiery Pisan in the letter to Christina. If God has total free will, argued Barbarini, nature cannot be the self-sufficient *prima donna* Galileo made her out to be. You cannot pin down the ultimate cause of some physical happening by studying its outcome. Even the cleverest scientist must never underestimate the multiplicity of God's reasons.

Despite his antipathy toward Galileo, Barbarini was a cultivated man, a poet of sorts, deeply read in the classics, fond of the arts, and fascinated by the new science. He respected the originality of Galileo's mind and generally regarded him as safe. When the two met, Urban, as pope, belabored the issue of God's omnipotence. He felt that Copernicus's theory could be no threat to the Church as long as it was unproven, and Urban believed it never would be proven, even by Galileo. Urban's main concern was to protect the integrity of the Bible, especially in the face of Protestant jibes that Catholics were minimizing its importance.

According to notes of the meeting taken by Agostino Oreggi, later the pope's personal theologian, Galileo, when taxed with Barbarini's criticisms, put on a devout face and said nothing. Oreggi wrote that by staying silent, Galileo, who when at ease with his friends was apt to smile at the notion that scientific theories are only human inventions, "showed that no less praiseworthy than the greatness of his mind was his pious disposition."

In the end, Galileo's literary gifts, his sharp pen and talent for satire, were his undoing. The Catholic Church was remarkably patient with him, considering his vanity and his prickly disposition. He was given permission to write on the Copernican and Ptolemaic systems of the heavens, as long as he did not explicitly take sides. The book that resulted was *Dialogue Concerning the Two Chief World Systems*. Highly readable, its tone a sort of refined scorn, the *Dialogue* is in the form of a conversation in a Venetian palace on the Grand Canal by three philosophers: Salviati, a mouthpiece for the author and Copernicus; Sagredo, an ostensibly neutral referee; and Simplicio, Italian for "Simpleton," a pedestrian defender of the old astronomy of Ptolemy and a butt for Salviati's wit. The Vatican was not amused.

Urban had been assured the book would include his stipulation that God's powers, being infinite, are unknowable to humans in their totality. Instead, Galileo had incorporated Urban's view on omnipotence, but in such a way as to deliver a crushing, mocking snub to his former friend. It was condensed into a few sentences, placed near the end of the dialogue, and, as a final slur, put in the mouth of Simplicio, the literal-minded, mental weakling of the trio. Simplicio lauds Galileo's theory of the motion of ocean tides, based on the Copernican system of a revolving Earth, but he declares it not true in light of a "more solid doctrine that I once heard from a most eminent and learned person, and before which one must fall silent," an oblique allusion to the meeting between Barbarini and Galileo. He adds that "God in his infinite power" could have made the tides move in other ways "which are unthinkable to our minds." It would be excessive boldness, Simplicio goes on, "for anyone to limit and restrict the Divine power and wisdom to some particular fancy of his own." By this device Galileo, with thinly veiled derision, expressed his private disdain for the idea that science, being finite knowledge, must always take second place to God's infinite and awesome power and freedom of will.

History does not disclose exactly what Urban had to say about the joke at his expense, and at the expense of his cherished doctrine, perpetrated in the *Dialogue*. But "someone," two leading Galileo scholars have recently concluded, "murmured in the pontiff's ear that he was being ridiculed." Whether because of the personal affront to Barbarini or his general blasphemy, Galileo was called to Rome to face the Inquisition and stood trial. In June 1633, *Dialogue Concerning the Two Chief World Systems* was condemned and Galileo was sentenced to house arrest.

Galileo did not pretend that science had all the answers. Mysteries remained, and probably would remain into the indefinite future. Amid those mysteries, there was room perhaps for God to work in ways science could not guess at. His omnipotence would ensure that those ways might not be what we expect. Nature has "abstruse" though not necessarily unfathomable modes of operation, which means that, in astronomy, new discoveries, though they give us godlike knowledge of one particular case, do not necessarily lessen the number of puzzles still remaining and may actually increase them. In the letter to Christina, Galileo wrote that the open book of the heavens contains more wonders than a mere spectacle of the Sun and stars, visible to "brutes and the vulgar." In it are couched "mysteries so profound and concepts so sublime that the vigils, labours and studies of hundreds upon hundreds of the most acute minds have still not pierced them, even after continual investigations for thousands of years."

In the Galileo affair, both sides were aware of the imperative claims of mystery, and the profound nature of the link between mystery and the sacred. With Newton it was different. The author of the *Principia* was not overfond of mysteries. For the most part he gave God specific, clearly defined tasks to perform: keep the planets in their stable orbits, prevent the system from imploding, guide the trajectory of comets. But Newton wanted God to be more than that, *do* more than that, know more than that. Straining against his desire to glorify the Almighty was his inclination to domesticate him. Here was the dilemma of his scientific investigations: they made the creation look like a piece of clockwork, and yet they could not be divorced from Newton's intense belief in a God who not only made the world but sustains it at all times, preventing its collapse. That left room for plenty of mysteries, for speculation as to what was due to the efficient ticking of the clockwork and what to the direct interference of the Creator-Governor.

Uncertainty, lurking at the core of Newton's system, never quite went away. It remained, a standing temptation to theorize about the religious implications of what was ostensibly a demystifying project. In the case of gravity, Newton never finally made up his mind as to what exactly it was, and clergymen made haste to exploit such indecision. It was an enticing gap in the supposedly airtight system, all the more so because it had to do with the absolutely central question of divine maintenance as opposed to divine creation. William Whiston, a prominent Newton disciple and a vehement anti-Trinitarian, wrote in *A New Theory of the Earth* in 1708 that "'Tis now evident, that Gravity, the most mechanical affection of Bodies, and which seems most natural, depends entirely on the constant and efficacious, and if you will, the supernatural and miraculous Influence of Almighty God."

At the time, there were four possible explanations of gravity, all left open. Was it an ether composed of fine-spun matter? A gossamer medium like light, through which divine activity flows? An effect of mysterious "active principles," not inherent in matter and akin to those proposed by the alchemists? Or simply an invisible and nonmaterial manifestation of God's preserving hand? Newton most definitely did not want to endorse a fifth possibility, that gravity is inherent in matter. Such a notion would undermine the power of the Lord of Dominion, introducing a force competing with that incomparable source of all that is. In a letter to his friend Richard Bentley, Newton wrote: "Gravity must be caused by an agent acting constantly according to certain laws, but whether this agent be material or immaterial, I have left to the consideration of my readers."

For more than half a century, Newton struggled to find a physical account of gravity. One theory was that it was produced by impacts of grains of an invisible ether, sweeping the planets in orbit around the Sun, according to laws set by God at the Creation. That put mechanism in charge and pushed God to the sidelines. For a time, during the 1670s, Newton seemed to reject this idea. His allies, Bentley and Clarke, confident that Newton would back them up, used gravity as evidence of the interaction between God and the physical universe. But in a London periodical, the *Newsletter* of December 19, 1717, it was reported that "Sir Isaac Newton has advanced something new in his latest edition of his *Opticks* which has surprised his physical and theological disciples." The surprise was that Newton had suddenly reversed his earlier objection to the ether as the mechanical cause of gravity. In the second English edition of the *Opticks* and in all later editions he proposed the existence of this elastic, rare, and subtle medium. But just because there is a medium, that does not identify the ultimate basis for this all-important force which holds the universe together.

The young Newton had stated explicitly that material bodies "cannot be truly understood independently of the Idea of God." In various forms and guises, this belief stayed with him to the end. Exactly what causes gravity became a crucial dilemma after the publication of the *Principia* in 1687. Newton's treatment of gravity as a new force flew in the face of the prevailing concept of "force" as impact, not of action at a distance. Newton juggled a number of different surmises, all the time insisting he was only describing what gravity does, not how it arises. He took great pains to assure his contemporaries that he was attempting to illuminate gravity's effects, not its cause, not its "sufficient reason." It was in this connection that he made his famous statement: "I feign no hypotheses." But while he did not come down openly on one side or another, he stuck with his partly theological repugnance toward the view that bodies possess an internal power of attraction. "I desired that you would not ascribe innate gravity to me," he told Bentley in 1693, calling the idea "so great an Absurdity, that I believe no Man who has in philosophical Matters a competent Faculty of thinking, can ever fall into it." Matter is passive and God is active. Matter needs to be put in motion and kept in motion by an active principle, which is nothing less than God's will. Of course, Newton allowed that God could use secondary agents to carry out this task. In any case, as he wrote in a draft corollary to Proposition VI of the *Principia*, "There exists an infinit and omnipresent spirit in which matter is moved according to mathematical laws."

Newton's firmly held belief in an "active principle" to explain gravity was a spur to his researches in alchemy, an apparent oddity in the career of a scientific mastermind for which scholars even today are unable to provide an explanation on which all can agree. In keeping with an activity dealing with the occult and strange, Newton's monumental labors over his furnaces and in his rooms in Trinity College were almost furtive. He guarded his experiments with a typically neurotic secrecy. He shared his methods only frugally, and with sympathetic scholars. Evidently, alchemy offered him further insights into what he found most elusive in his theory of gravity and strengthened his belief that mechanism was not the whole story. Newton regarded true alchemy, which aimed to induce the processes of "growth and life" similar to those in plants into metals, as "more subtile and noble" than merely imitating mechanical changes in nature, which he dubbed "vulgar chymistry."

In Newton's mind, true alchemy used nature's own mediator, a "vegetable spirit," a microscopic particle which diffused into matter and started the processes of growth and activity. The goal was to bring this spirit under control, making the alchemist master of the secret world of organic life. If he was successful, God's agent would become his agent, a heady prospect, and one an alchemist is extremely reluctant to talk about to just anybody. Newton told Robert Boyle some of his ideas about the vegetable spirit in a letter of 1679, but warned that they were not "fit to be communicated to others." He believed the story of the Creation in Genesis was an encoded account of an alchemical process. God the alchemist made the world by organizing matter via the vegetable spirit, and the human adept was trying to accomplish exactly that in his laboratory. Newton himself compared his work with "bringing forth the beginning out of black chaos and its first matter through the separation of the elements and the illumination of matter."

The active principles at work in gravity, Newton decided, were of the same kind as the active principles of the vegetable spirit. Just as they produced and conserved motion, so too they caused organic growth in plants. In both cases, they gave life to lifeless matter. We meet with very little motion in the world, he said, "besides what is owing to those active Principles. And if it were not for these Principles, the Bodies of the Earth, Planets, Comets, Sun, and all things in them would grow cold and freeze, and become inactive Masses; and all Putrefaction, Generation, Vegetation and Life would cease, and the Planets and Comets would not remain in their Orbits." Active principles were another manifestation of God's sovereign power and will.

It seems that Newton exploited the idea of occult powers while keeping in mind the possibility that one day they might be explained in rational terms. That added to the confusion over what God does or does not do in the everyday running of the universe. Before the Big Bang of the scientific revolution there had been a tradition of "natural magic," which held that the world is full of powers that are occult but at the same time are part of nature. John Henry, a historian of science, notes that practitioners of this kind of magic aimed to identify and put to work such quasi-supernatural forces, so as to manipulate the organic and inorganic world. The idea was to combine unlike things and separate like things in order to bring off "what the vulgar call miracles." Newton had the leading manual, *Natural Magick* by Giambattista della Porta, 1558, in his library.

Some of today's thinkers propose that God could enter more easily into our physical world as portrayed by modern theories of uncertainty than into Newton's universe, because his lacks a sufficient degree of "strangeness." But for Newton the alchemist no less than Newton the physicist the world was indeed strange, and the possibilities for manipulating that strangeness had only begun to be explored. In natural magic, it is a question of what actives work on what passives to obtain a given outcome. But the causes are mysterious because there is no explicit test by which to demonstrate them. Newton's library contained 1,752 books, of which as many as 170 dealt with magical topics. John Henry argues that Newton's concept of force, the key novelty of his science, was derived from ideas of occult powers in the natural magic tradition. In a paper entitled *An Hypothesis Explaining the Properties of Light*, read to the Royal Society in 1675, Newton suggested that all bodies may be made up of "certain aetherial spirits or vapours," condensed in different degrees and "wrought into various forms." This power of assembly and organization with both light and ether as mediators was set in train by God with the instruction to "increase and multiply."

But the same idea is to be found in Newton's much more secret accounts of his researches in alchemy. A year earlier, in one such report, he had stated that "the aether is but a vehicle to some more active spirit and the bodies may be concreted of both together, they may imbibe aether as well as air in generation and in the aether the spirit is entangled. This spirit perhaps is the body of light because both have a prodigious active principle and both are perpetual workers." Newton surmised—he who did not "feign hypotheses"—that all the operations of nature might be derived from unknown forces of attraction and repulsion that were at work in the realm of alchemy. In a draft preface to the *Principia*, he said he "suspected" that everything in the universe depends on these two forces; but just how they work is a mystery. "Life and will are active principles," he wrote in a draft of Query 23 for the 1705 edition of the *Opticks*. "Laws of motion arising from life or will may be of universal extent."

Newton was more cautious than his admirers about explaining the crucial forces that wake dead matter into activity, whether due to God directly or to a mediator working on his behalf. But certainly it became possible to think of God and of his relation to the world in new ways, some of them abhorrent to Newton's deeply held beliefs. Popular writers on religion took up the question of gravity, treating it almost as theologians used the idea of God's omnipotence, even inviting a comparison with the doctrine of grace, bestowed on an undeserving world as a gift. Without the input of divine energy, the physical world would be without motion, without purpose, without meaning, just as human beings are not redeemed unless they receive the gift of grace. Humans need not have been saved, and the world need not have been put in motion; they are saved, and it is in motion—thanks only to the will of God.

During the eighteenth century a picture of God as radically, uncompromisingly sovereign, a deity that perhaps Luther and Calvin would have looked askance at, owed more than a little to the ambiguity Newton left open in his mechanical-alchemical theories of nature. The historian of religion Gary Deason thinks the God of the new science overwhelmed the powers of nature to such an extent that nature, now too passive to assert itself, "points to the necessity of God." He was being given so much to do keeping up his creation, and matter so little to do, that the idea of him could not stay as it was. Luther and Calvin wanted a supremely sovereign deity for a special reason: they needed to press home the message that God alone saves, while humans in their puny scamblings after merit can accomplish nothing. This was not a philosophy of nature but a key to Reformation theology. A God with total power over nature meant his creatures need not worry unduly about being struck by lightning or buried in an avalanche. If they were, they could rest easy in the thought that there was a good, if inscrutable reason. This helped to counter superstition. It also made rational questioning of the doctrine of predestination fruitless. God's utter supremacy and majesty put him beyond reach of human reason. Some Puritans were actually ready to suffer eternal damnation if that increased God's glory. And Calvin's fixation on divine omnipotence in time served to promote the idea of equality: all persons are equally undeserving in the light of divine perfection.

The Newtonians likewise insisted on an omnipotent Creator, but for different reasons. They had a high respect for the laws of nature, leading them to conclude that these would be overridden only on rare and special occasions, so that while God keeps the operations of the physical universe running smoothly as a sort of resident engineer, he does not go in for trivial acts of intervention. He oversees the long-term career of the world, its "general providence" and grand enterprises, its overall goals. The God that Newton's followers constructed from his science was a cosmic governor, a commander in chief, ruler of the universe, not a tinkerer or arbitrary interferer.

Luther and Calvin, working in crisis conditions, saw it as an urgent task to correct a mistaken view that had got the doctrine of grace back to front. Medievals supposed we are partners with God in the work of salvation. The Reformers turned that idea upside down. If being saved depends in the slightest degree on our own efforts, that detracts from God's adequacy. It steals some of his thunder, elevating the human role at the expense of the divine. The Newtonians applied this same principle to the physical world. Matter was inert, passive, without a mind of its own. It did not collaborate with its creator. And the agency via which God worked, the physical equivalent of divine grace, was Newton's gravity.

As a result, Deason surmises, the character of God underwent an alteration. "As the key to the meaning and structure of the new mechanical world, gravity became the mark of power and grace. . . . ," he writes. "The sovereign Redeemer of Luther and Calvin became the sovereign Ruler of the world machine. The Reformers' search for assurance of salvation gave way to the assurance of scientific explanation."

CHAPTER NINE

"One of History's Great Losers"

God being therefore hidden, any religion which does
not say that God is hidden is not true.

　　　—BLAISE PASCAL

In the Incarnation, God became veritably a secret:
He is made known as the Unknown, speaking in eternal
silence; He protects himself from every intimate
companionship and from all the impertinence of
religion. He becomes a scandal to the Jews and to the
Greeks foolishness.

　　　—KARL BARTH

M ORE THAN ONE KIND OF GOD emerged from the turmoil of the
Reformation and the radical turn to modernity spurred on by the scientific revolution. Newton's intense desire to "get God right" was part and
parcel of his scientific theories and his conjectures, public and private, as
to the nature of God's universe. It was implausible that he and his opponents could be talking about the same deity, and the history of religion
then and afterwards tended to be written as if there were several. Something similar happened to ideas about Jesus in the later twentieth century,
when different cultural cliques adopted different images of a Christ who
turned out to be multiply acclimatized to all kinds of exotic interests.

The uniformity of science's God was broken in dramatic fashion by a celebrated exchange of letters between two of Newton's contemporaries, Samuel Clarke and Leibniz. Newton harbored a deep suspicion that Leibniz had stolen the idea for calculus from him. Because of that bad feeling, the tone of the letters is offensive in places and bilious in others. Leibniz held that God is both intellect and will, but intellect has the larger role. God created the "best of all possible worlds"—an idea famously exploded by Voltaire—and therefore he needed to understand all the other possible worlds. Such a philosophy tends to diminish God's activity in the actual world, since he is intelligent enough to have chosen the best world, which, being the best, presumably would not need day-to-day tinkering once it got started. To this Clarke replied that Leibniz had missed the whole point of Newton's argument. The idea that the world is a great machine running forever without the intervention of its Author "is the notion of materialism and fate." It hints at a universe which never had a beginning, and therefore no creator. If a king had a kingdom where everything happened without his government, "it would be to him merely a nominal kingdom; nor would he in reality deserve at all the title of king or governor."

In fact, Leibniz was not overly fond of kings who ruled with absolute power. That included Louis XIV, the Sun King, epitome of the absolute monarch, addicted to glory and majesty. He thought Louis, with his expansion-happy foreign policy, was a menace to good order in Europe. Leibniz was unfriendly to theories of nature which stressed God's dominion and unfettered will, feeling that they tended to give dangerous potentates like Louis the wrong idea. He highlighted the rational element in godhood; justice would prevail in the world "not so much by the Will as by the Benevolence and Wisdom of the Omniscient."

Leibniz was born in Leipzig soon after the end of the Thirty Years' War, the son of a philosophy teacher. He studied law and then took a position with the archbishop of Mainz, on whose behalf he attempted to hold secret talks with King Louis, then threatening aggression against Holland, to persuade him into making war against the infidel in Egypt and Constantinople instead of the faithful in Europe. He was rebuffed, being curtly informed that "holy wars went out with *Saint* Louis." Bertrand Russell, who clearly had no love for Leibniz, accused him of sucking up to princes, in spite of his professed distaste for the aggrandizement of power. Leibniz was somewhat tight with money, a quirk Russell went out of his way to mention. As a librarian at the court of Hanover, Leibniz would donate as a wedding "present" to a young lady getting married a handwritten manual of advice, including a warning not to "give up washing" when she was safely esconced as a bride. Others say he was a gentle and friendly person who enjoyed the company of children.

What set Leibniz apart from Newton was his theory that the world consists of an infinite number of substances, called "monads," which can be thought of as physical units but are really more like souls. Each monad, he said, mirrors the entire universe. God gave monads this mirroring property, and he ordained, at the Creation, a "preestablished harmony" whereby all the monads are kept in a perfect relation with each other, without cause and effect. If Newton's universe was like a gigantic clock in need of repairs now and then, Leibniz's was a consortium of innumerable clocks, all faultlessly synchronized to strike at the same instant. God stands outside this wonderfully concordant system because it has no need of him. Leibniz complained that Newton's demeaning idea of God as being required to adjust the cosmic clockwork from time to time was a reason for the decline in religion in England. Such a jibe was all the more hurtful since at that time many public clocks were bulky iron or bronze affairs built by blacksmiths, so inaccurate early on that they lost or gained up to fifteen minutes a day. By Newton's time, this discrepancy had been reduced to about seven minutes. A keeper or governor had to wind a clock twice a day and reset the hands. In a sneering remark, calculated to get under the Newtonians' skin, Leibniz said: "Sir Isaac Newton and his followers have a very odd opinion concerning the work of God. According to their doctrine, God Almighty wants to wind up his watch from time to time, otherwise it would cease to move. He had not, it seems, sufficient foresight to make it a perpetual motion. Nay, the machine of God's making is so imperfect, according to these gentlemen, that he is obliged to clean it now and then by an extraordinary concourse, and even to mend it as a clockmaker mends his work."

We should remember that such statements were made in the years 1715 and 1716, when Leibniz was ill, riddled with gout, and disappointed that George I, on taking the throne of England, had rebuffed his overtures for a post at the English court. His long absences from Hanover and his quarrel with Newton over the calculus may have shipwrecked his chances.

Leibniz makes it clear that his God is not the same as Newton's. He is less willful, less wedded to power, and he seems to be more intelligent. A God whose intellect is his most important faculty is likely to make a universe of greater harmony, greater integrity, and greater beauty than a God of will. That latter type of God is a little more unpredictable, a shade less rational than the fastidious one of Leibniz's scheme. There is an aesthetic quality to Leibniz's divinity, and by contrast an intimation almost of crudity, of philistinism, in Newton's God, who lives from day to day, keeping the world on track. Leibniz repeatedly uses the word "beauty" in his rebuttal of Newton's theology. If God has to step in periodically, that means he must have second thoughts about his creation. "No," Leibniz counters. "God has foreseen everything; he has provided a remedy for every thing beforehand; there is in his works a harmony, a beauty, already preestablished."

This was a sensitive issue. Margaret Jacob has pointed out that Newton's firm belief in divine action and regulation of nature was the reverse side of the coin of his occasional visions of disorder and even chaos in the universe. In the first edition of the *Principia* in 1687, Newton suggested he now doubted the endless, uninterrupted harmony of the various parts of the cosmos. The Earth could grow larger until the balance of the universe was upset. Perhaps he was thinking of an ultimate apocalypse, or of God's punishment for a sinful world. If humans defy the will of God, then physical disorder may follow. Samuel Clarke tells Leibniz: "'Tis in the frame of the world, as in the frame of man's body: the wisdom of God does not consist, in making the present frame of either of them eternal, but to last as long as he thought fit."

Where does God reside: inside the world or outside? Leibniz called him a supramundane intelligence, author of harmonies, perfect designer of a world perfectly in order. The actual world is the best of all possible worlds, but God stands outside it, and outside all possible worlds. Included in the notion of God's "perfection" is the stipulation that he should stay removed from the affairs of the physical universe he made. The world wants to be influenced by its creator, but Leibniz seems to be saying it *could* function successfully on its own. That is because God is more intelligent than Newton and his admirers, who emphasize divine power, seem to suppose. Samuel Clarke amended the term "supramundane" to "omnipresent"—present inside *and* outside the world.

Newton wanted God to be involved in the world in whatever shape or form and suggested entry points for him. In his alchemical researches he gave a starring role to the active, animating agent in matter, which he called by the code name "magnesia." Magnesia breaks up into anarchy the bits of matter and then arranges them into a new configuration. In this way, base metal could become gold. Newton floated the idea that the world is a quasi-animal, inhaling and exhaling a breath of ether for its daily refreshment. This breath is an agent of God's will, operating in the universe, a kind of "material soul." It is a mediator between God's otherness and our own realm. Active principles, at work in alchemical processes, were one of the points of entry for the divine into common nature. They were halfway between the celestial and the mundane. "Couriers from the beyond" is how the Newton scholar Betty Jo Dobbs described them. Wearing his scientist hat, Newton is on record as stating categorically that there are no mediators. But in his treatises on alchemy and theology it is a different story. God is not the soul of the world, but the alchemical spirit is his "contractor." That is flat contrary to Leibniz's theology of a deity who needs no intermediaries or "agents" to keep the universe in harmonious operation.

Leibniz also studied alchemical texts, but not with the same intense, disinterested quest for truth. A polymath, he was intrigued by a hundred byways of learning and ideas for new inventions: a submarine, a power saw, a binary code that centuries later became the basis of the electronic computer. Alchemy was just another interest. In the crude sense of turning base metal into gold, it might make him financially secure. Famously close with money, Leibniz during his tenure at the Hanover court invited several traveling alchemists to demonstrate their art and actually invested money in one of them, Jonathan Craft, funding his experiments.

It is understandable that two so very opposite theories of the relation of God to his world should have emerged. We are faced with a bewildering variety of attitudes and moods in the Old Testament's depiction of Yahweh, leading some scholars to wonder if the writers of these texts were not blowing off steam, making God responsible for hardships they thought were undeserved. An extreme, uncompromising monotheism, which Newton strongly approved, may veer away into one of two quite different directions. One is an inclination, as God becomes increasingly transcendent, to internalize the devotion that was once given to a "real" Supreme Being. The other is a propensity for domesticating and humanizing a deity so abstract and mysterious he is at risk of disappearing altogether.

The theologian Richard Wentz thinks the domestication of the divine began in earnest with the late medieval debate as to whether God is primarily will or chiefly intellect. This dilemma could hardly have arisen in early biblical times, when God was a sort of untameable force, a whirlwind of creative energy, whose appearances were often accompanied by storms and bolts of lightning, as well as generous quantities of smoke. When he summoned Moses up to the mountain, the mountain, as well as the Israelites, trembled. He was a powerful presence, never still, always doing things. But when a distinction is made between will and intellect, as Clarke and Leibniz certainly did in their exchange of letters, God becomes an object, to be classified as one or the other, as one would sort a specimen into a given class or category. If we maintain that God is primarily will, then we increase God's separation from a human nature that is increasingly based on reason, especially scientific reason. It places a tremendous emphasis on "belief," which was almost a fixation at the time of Luther and Calvin. According to the thirteenth-century Scholastic thinker Duns Scotus, belief is not first and foremost a movement of the intellect, but rather an act of will. Will, he writes, can be stubborn, oblivious to facts, hostile to other and different expressions of will. This hostility can manifest itself in a number of ways. Scotus thought that sin was the result of a rebellious will in human beings. Reason may arrive at an idea of the character of God—a sort of intellectual polytheism—quite contrary to that of a willed belief. A merely cerebral reading of the Old Testament might fasten on his "bad" qualities, his anger and alterations of mood, his inexplicably shabby treatment of Job. That could lead to the very dangerous heresy that the reality of God depends on the willingness of human beings to believe in him.

A case has been made that Newton's near obsession with God's dominion strengthened the hand of certain sections of society in advancing their position. Among these were liberal clerics of the Church of England known as Latitudinarians, moderate men with a vested interest in the status quo, loyal, at least on the surface, to the king as head of the Church, enemies of all the divisive elements that had led to a bloody civil war and the execution of a sovereign. Now they were intent on preserving the crown and maintaining an orderly society. Newtonian science-cum-theology was just the right prescription for a new world fit for Anglican clergymen and the rising business class to prosper in. The Newtonian cosmos, guided by providence and tamed by the artifice of mathematics, was a model for an orderly and prosperous nation. Newton himself had written that "The whole world natural consisting of heaven and earth signifies the whole world politic consisting of thrones and people." Providence would guarantee that the new political and social regime proceeds on an even keel, just as it sustained an orderly physical world, intelligible to scientific inquiry.

Clarke proposed that "a frequent and habitual contemplating of the infinitely excellent perfections of the all-mighty creator and the all-wise governor of the world" tends to foster the social virtues. And those virtues in turn support the kind of society in which commerce and the capitalist enterprise can thrive. It was a case of Newton to the rescue for the moderates, who needed an infusion of social calm and consensus, conditions indispensable for the proper working of a new, competitive market system. Without a stable framework of order, of a strong central government, the delicate balance of constraint and freedom that kept the system doing what it was supposed to do could break down. Margaret Jacob argues that Newton's universe, stable and harmonious, was peculiarly suited to a new ideology where "the necessities of the marketplace presented no contradictions to the dictates of natural religion." Latitudinarians blithely wedded the Newtonian system of godly providence and human mastery of nature to the rising commercial prosperity of England.

Churchmen who lectured in the series endowed by Robert Boyle had no inhibitions in this respect. In 1711 William Derham, in lectures entitled *Physico-Theology: or, A Demonstration of the Being and Attributes of God, from the Works of Creation*, went so far as to apply Newton's proof of design in the universe to England's project of building an empire overseas. Developing trade links with China, said Derham, aided the spread of missionary Christianity. That was as much a part of God's plan as the maintenance of the heavenly bodies in their courses. "Thus the wise

Governour of the Worlde, hath taken Care for the Dispatch of Business," Derham announced. In fact, providence had also ensured that commerce, while looming large, did not monopolize life so completely as to leave no time for worship of the Author of such an ordered world: "But then as too long Engagement about worldly Matters would take off Mens Minds from God and Divine Matters," Derham advised, "so by this Reservation of every Seventh Day, that great Inconvenience is prevented also." Another Boyle lecturer, John Harris, in 1698 had claimed that the virtues of religion—liberal Protestant presumably—"do naturally and essentially conduce to the Well-being and Happiness of Mankind, to the mutual Support of Society and Commerce, and to the Ease, Peace, and Quiet of all Governments and Communities." In time this laid the Church open to the accusation of conniving in the miseries of the industrial revolution. "William Blake was not being simply rabid when he assaulted Newton as the symbol of a market society, based on technology and empire, which oppressed him," Jacob writes.

As it turned out, science eventually owed as much to the Latitudinarians as they owed to science. Their sermons championed the Newtonian revolution, the new "mechanical philosophy," contra Leibniz, making it palatable to an audience whose level of education might otherwise have inclined them to skepticism. A distinction began to be made between "natural" and "revealed" religion, the truth of the first being discoverable by ordinary human reason, the second coming from elsewhere, direct from God or via other channels, making each particular religion different from the others. In principle science did not make use of messages from "elsewhere," but Newton was a special case: nature's secrets can be prized out by inspection and deduction, subject of course to the X-factor, the will of God.

Latitudinarians, "men of wide swallow," in the famous phrase of the polemical Edward Fowler, were intent on making their doctrines imprecise enough to create a theological big tent, sufficiently roomy to accommodate dissenters as long as they were not actually heretics. Their tolerance was even extended to Quakers, but not to Catholics. The idea was to permit some latitude—which meant tolerating a measure of uncertainty—to sects once regarded as beyond the Anglican pale. Earlier in the seventeenth century the Puritans, men of rather small swallow, had kicked off a civil war and put a king to death. They were blamed for the havoc that accompanied this turning "the world upside down." Newton's solid, stable cosmology, underwritten by God's will, might help to keep it right side up.

These liberal clerics aimed to give as little offense as possible to as many people as possible. They were hostile to the theory of predestination, which was a trademark of Puritanism, and Puritanism, as Charles I's archbishop had said, was the root "of all rebellion and disobedient intractableness and all schism and sauciness in the country." At the other end of the spectrum, papists were definitely unwelcome. James II, who succeeded his brother Charles II as king, learned that lesson the hard way. When he tried to make Catholicism acceptable in England, the anti-Catholics removed him from the throne. Almost at once, the Toleration Act of 1689 underwrote the new climate of vagueness in religion which was just to the Latitudinarians' taste. It was expected to calm the country down, and with its stress on decent behavior bring in a period of social tranquility.

The trouble was that many admirers and adherents of geniuses like Newton and the philosopher John Locke took their ideas further than their authors would have dared. Locke, a friend of the latitude men, conceded that there are mysteries too "other" for us to grasp, but held that reason confirms God's existence with a cogency "equal to mathematical certainty." His devotees accepted his teachings on the power of reason but threw overboard his caveat that God can reveal truths which are outside its scope, and that faith in Jesus as mediator can rescue us from perdition even if our moral lives are not beyond reproach. The latitude men did not go to such extremes, but they altered the religious atmosphere in such a way that extremes became possible. The breadth of their toleration matched the expansiveness of their interests, which included "the world politick" as well as social theory. They were especially interested in ideas about how the state Church could secure its position. For them the universe was a machine, a wonderfully ingenious and orderly one, obeying the mathematical laws Locke said showed divine authorship; it was guided by the providential hand of God. Newton's universe, in the words of Margaret Jacob, allowed the latitude men "to imagine that nature was on their side; they could have laws of motion and keep God; spiritual forces could work in the universe; matter could be controlled and dominated by God and by men."

An irony of this situation was that moderate churchmen were not quite on the same wavelength as their congregations, many of them made up of prosperous Londoners. Thinkers like John Wilkins, Robert Boyle, John Tillotson, Isaac Barrow, thought they were defending the world against heterodoxy, from deviationists broadcasting the dangerous belief that God is *in* nature, or is more or less the same as nature. The important thing was to leave room for God to enter at crucial points of uncertainty without installing him as part of the material world, a precarious tightwire act. Once you believed that the whole Creation, in the words of Gerrard Winstanley, a London shopkeeper turned radical socialist, "is the clothing of God," the floodgates were open to all kinds of quasi-socialist notions about all men and women being equal in a God-filled universe; about priests and those in authority being superfluous, since everyone imbibed the same elixir of the divine. And if matter is spirit, then fringe pursuits like magic, alchemy, and astrology could be made almost respectable. Worse, in the eyes of Catholics, it would make redundant the doctrine that Christ is selectively present in the communion bread and wine, a sacrament managed by priests possessing indispensable power to transmute mere matter into the body and blood of Jesus.

In times of transition, new kinds of uncertainties are apt to surface. Religious people sometimes react by inventing otherworldly means of making the world more certain. At a moment when the Reformation was at risk of falling into disarray, Calvin insisted on the doctrine of predestination, designed to calm anxiety as to one's fate in the afterlife. That made the mediation of confessional priests surplus to requirements. But uncertainty can also be a catalyst for the arrival of new kinds of mediators and different ways of relating to the divine. When England was in transition from a monarchy to a commonwealth and back to kingly rule again; when people asked themselves, "Do we want a king or don't we?", a bewildering profusion of sects emerged, each with its own peculiar version of divine sovereignty and means of relating to it. It was a chaotic and socially driven form of polytheism. Different agendas, different Gods. Baptists, Quakers, Muggletonians, Seekers, Ranters, produced their own answers to the dilemmas thrown up by the turmoil of the times, upheavals that led Winstanley to say that the old world "is running up like parchment in the fire."

Christopher Hill called these eccentric spinoffs a "third culture," between the established Church and orthodox Puritanism. They blossomed at a time when, apart from unpredictable catastrophes of famine and plague, there had developed a harsher spirit of competition in trade and commerce, key factors in the development of an "Age of Power," which introduced a new element of insecurity. As business and commerce began to take hold of the English economy, people resorted to marginal aids to psychological comfort not approved by the Church. Leading politicians like Oliver Cromwell, Bulstrode Whitelocke, and Richard Overton, the Duke of Buckingham, consulted astrologers. And with the Protestant emphasis on the voice of the individual conscience, a space was made for God to enter unmediated into the minds of his elect, for example via dreams and visions. Winstanley had visions, as also did George Fox. Some people who talked to Fox in the mid-seventeenth century told him they relied greatly on dreams. Such a meeting of minds would eventually blossom into the Quaker movement.

One reaction to this uncertainty, bred by the turning of the world upside down, was simply to wait for God to come to the individual person. Mediators being out of fashion, the feeling at this time was that one should be quiet and pray patiently for illumination, one-on-one. Such was the attitude of a fringe movement called the Seekers, some of whose members later became Quakers. The Seekers were taught to "sit still, in submission and silence, waiting for the Lord to come and reveal himself." William Erbery, an army chaplain and a promoter of the Seekers, preached that there were no intermediaries and that by this means God would lead the prayerful ones "out of this confusion and Babylon." Erbery, whose wife and daughter became Quakers, recommended being completely alone, in "a wilderness condition, which with God is the most comfortable state." To attend church and worship with others was to commit "spiritual whoredom." He believed that God was "tumbling the earth upside down and tossing to and fro the government," making it impossible to attain any certainty.

In this unsettling atmosphere, where the habit of authority was giving way to the possibility of a personal avenue to God, the latitude men offered a piety that took seriously the voice of the individual conscience, but also called for an end to sectarian rivalries. Times of transition, they suggested, are complex and disruptive. But Newton's new picture of the cosmos, of God's creation, is *scientific* proof that beneath the confusion, behind the uncertainty, is the hand of the all-powerful *pantokrator*. That guiding spirit is more real, they felt, and more powerful, eventually, than the apparent disorder. Newton had suggested that physics, as opposed to biology, is better suited to demonstrate the inherent orderliness of the cosmos because it uses relatively few and quite simple universal laws. He told Richard Bentley: "When I wrote my treatise upon our Systeme I had an eye on such Principles as might work with considering men for the beliefe of a Deity and nothing can rejoyce me more than to find it usefull for that purpose." Newton's friends, men like Bentley and Samuel Clarke, went public in the pulpit and in print with their message that in the transition to a new age, when it was perfectly proper to pursue wealth and success, one must accept the existence of an immense cosmic order put in place by an exclusively monotheistic God, an interventionist God of will, and try to emulate it in the smaller worlds of society and politics. A liberal and tolerant Christianity would not be possible without this sheet anchor of constancy.

These men were impeccably serious and devout in their reinvention of Christianity. But the next generation of Newtonians were another story altogether. Considerable skill had been needed to sell the idea of an essentially mechanistic universe as a reason to believe in divinely imposed order and harmony without adopting Leibniz's scheme of a harmony inherent in matter. Now, as the eighteenth century proceeded, the participation of the deity was increasingly minimized. The "designer" role of God, as opposed to his active guidance, was expressed and supported by new institutions like the Freemasons, officially installed in London in 1717. A leading light in the organization was one of Newton's friends and fervent admirers, a French refugee named Théophile Desaguliers, an experimenter and teacher of Newtonian science in London. Under the leadership of Newtonians, Margaret Jacob has shown, this secret society became a force for religious tolerance and loyalty to the crown. The God of Freemasonry was only half of Newton's God, however. He was referred to as "the Grand Architect," a Maker but hardly an intervener in the particulars of his creation. Members wanted a roomy sort of religion that could accommodate a broad spectrum of creeds and doctrines. Such desires required a different sort of God. The once-dreaded pantheists—and it had been thought that a clockwork universe was an antidote to pantheism—jostled elbows with the Newtonians and an assortment of materialists. Some were Deists, whose creed entailed little more than the proposition that simply because God exists, you should behave yourself.

In fact, it was the well-behaved but not necessarily "caring" cosmos that caught the imagination of post-Newtonian intellectuals rather than Newton's views on the nature of God or the purification of the Christian religion. "The assimilation of Newtonian science into Western thought," Jacob writes, "produced the first generation of European thinkers for whom faith in the order of the universe proved more satisfying than faith in doctrines, creeds and clerical authority." It was this shunting aside of theology, to which so much effort and study had been given, replacing the idea of an active, involved deity who not only created "this beautiful, pointless machine" but watches over it and guides its destiny, that led Betty Jo Dobbs, one of the twentieth century's most respected scholars, to decide that in matters to which Newton had devoted his most ardent and intensive efforts, the verdict of posterity must be that he was a failure. There is a perverse tendency for theological doctrines, over time and in the hands of dedicated followers, to acquiesce to precisely what the original doctrine went to great pains to avoid. Newton's system hinged on a never-sleeping providence, but the idea of providence was already in the early stages of being secularized. As the discoverer of the theory of gravitation, Newton had won his huge celebrity, in the popular mind, as the man who showed that matter might be able to move itself. In time, thinkers who subscribed to the idea of an absentee God recruited Newton for their arguments, something that would have scandalized him had he known.

There might be some merit, Dobbs announced in a lecture to the History of Science Society shortly before her death, "in evaluating Newton in a different way, not as one of history's all-time winners, nor as the First Mover of modern science, nor as the Final Cause of *the* Scientific Revolution, but as one of history's great losers, a loser in a titanic battle between the forces of religion and the forces of irreligion. Perhaps he is not less a hero from that perspective, but a hero of a different sort, rather more like Roland at Ronceval, crushed by overwhelming odds in the rearguard, than a peerless leader of the vanguard."

CHAPTER TEN

A Cool Mediator

What I have learned about God in a lifetime of seeking
is that first, God must be sought in the light, and that
second, God does not have to be found.

—JOAN CHITTISTER

There's this great image of Moses hiding in the rock
and God goes by and Moses only sees God's back, and I
remember sitting in one Saturday morning for Sabbath
services, and saying that God's back is science.

—COMPUTER SCIENTIST MITCHELL P. MARCUS

THE IDEA OF SCIENCE as a mediator, taking over some of the functions formerly belonging to priests and supernatural beings, was one of
the causes of an institution which for a time played a dominant role in a
culture increasingly bent on improving the material condition of
mankind. This was the Royal Society, formed in Britain in the "wonderful pacifick year" of 1660. In its origins, the Society was well disposed
toward the orthodox in religion as well as in politics, and unfriendly to
effusively emotional types of piety professing to have exclusive and
unmediated access to God. Both were apt to destabilize the status quo.
Only seven years after its inauguration, a "history" of the Society was
published by Thomas Sprat, bishop of Rochester, a piece of writing we

are warned to examine carefully for hidden motives and easily over-looked partialities.

Sprat used the opportunity of an authorized account to let fly some barbs at occultists and those who dabbled in private revelations and various species of the supernatural. Science was a formidable newcomer as a secular power, but true religion need have no fear of it, he assured his readers. Only bogus prophets and those making fanciful claims to be divinely inspired had better look out. He makes it clear that the new science is a bridge between our minds and the mind of God. Science "is a religion," Sprat states boldly, a fact "which is confirmed by the unanimous agreement of all sorts of Worships." He recognizes that there is an immense distance between the human and the divine intellect. We are unable to imitate the slightest effects of nature, though fresh insights into nature may lead us to adore the godhead through the beauty, contrivance, and order of his works. But, and this is the crux of the matter, at least the scientist is more attuned to God's technical genius than the devout but untutored lay person. The prayers and praise of the scientist are more acceptable to God than "the blind applause of the ignorant." It is clear that science as mediator is available only to a select few, rather as Calvin's predestination would pick and choose its fortunate elect.

And science is above all a *cool* mediator, as cool as the weather in fog-shrouded England. As Sprat saw the situation in England at that time, it was one in which the excitable and heated had gained ascendancy over the serene and sober in matters of religion. He wrote as if this were a first-class crisis, putting the very survival of decent society at risk. Many scientists, he thought, suffered from a creeping skepticism and a neglect of worship, and it may be that they were driven to these regrettable lapses by the profligate excesses of people who had abandoned even the bare minimum of mediators approved by the Protestant code. "The infinite pretences to *Inspiration*, and *immediate Communion with God*, that have abounded in this *Age*," Sprat complained, "have carry'd several men of wit so far, as to reject the whole matter, who would not have bin so exorbitant, if the others had kept within more moderate bounds." The remedy is not to return to medieval mystery mongering, but rather to quell "the violence of *spiritual madness*."

Experimental science, Sprat said, had been hindered by "the great a-do which has been made, in raising, and confirming, and returning to many different Sects, and opinions of the *Christian Faith*." The wit of man "has been profusely poured out on *Religion*, which needed not its help, and which was only thereby more tempestuous while it might have been more fruitfully spent, on some parts of Philosophy, which have been hitherto barren." Sprat wrote in the language of emergency, not in the steady voice of an author confident that religion is on an even keel. The times reminded him of the period leading up to the collapse of the gods of antiquity, when people paid lip service to religion in public but had quite a different attitude in private. When not taking part in assemblies, they had no use for observances at all. Sprat took a contrarian view of the waning of genuine religion in England in his day. Society had been brought to a point where its very soul was drifting into the shallows. But that was not chiefly on account of an epidemic of the secular sins of greed, adultery, crime, sharp practice, cruelty, and carnal pleasures of the flesh. Just as pernicious were spiritual vices: the puffed-up sense of having exclusive access to God, the doctrinal fanaticism of fringe sects, the hysteria and orgiastic exhibitionism, the notion that a person full of the inner light may behave disgracefully since "to the pure all things are pure." If it had been an age of sexual license and reckless gratifying of the baser appetites, then the deep mysteries of the Christian religion could properly have been brought into play, purging such grossness. But the kind of religious distemper running rampant in Sprat's time was "subtil, refin'd and Enthusiastical." And it required a cure that used rational argument and plain sensible explanations.

A distaste for theological bickering had been one of the motivations for starting a scientific body of this sort. The Society's cradle and birthing place was Wadham College, Oxford, where meetings of "Vertuous and Learned Men" took place. Those men included Robert Boyle, Seth Ward, who became bishop of Exeter, Wilkins, Sir William Petty, Christopher Wren. According to Sprat, they desired only "the satisfaction of breathing a freer air, and of conversing in quiet with one another, without being engag'd in the passions, and madness, of that dismal Age." They were "invincibly armed against all the inchantments of *Enthusiasm*." Away from the obsession with religious doctrine that was tearing the country apart, they could discuss calmly and in privacy matters scientific, which still, to them, had a sacred dimension. Such spiritual frenzies, said Sprat, as "did then rule, can never stand long, before a cleer, and a deep skill in Nature."

If any nation was suited to lower the spiritual temperature that soared to fever pitch between the reign of Charles I and that of Charles II, it was the English. That was Sprat's opinion. Only the English have a gift for speaking and thinking plainly, an antidote to the disease of mystical agitation. It is to their credit that they have "commonly an unaffected sincerity; that they love to deliver their minds with a sound simplicity; that they have the middle qualities, between the reserved subtle southern, and the rough unhewn northern people: that they are not extreamly prone to speak, that they are more concern'd, what others will think of the strength, than of the fineness of what they say; and that a universal modesty possesses them." Of course, foreigners might look at these same sterling qualities and call the English slow, silent, sullen, and marked by a "melancholy dumpishness."

Without a blush, Sprat calls the English a chosen people; chosen to unveil the secrets of God's creation. Nature will disclose more of her inner workings to the English than to others, since she has already provided them with a "Genius" for discovering them. And what goes for England goes too for the Church of England, since it is the most reasonable Church, the one that befits a time of more settled and tranquil politics. Foreigners, trapped in a papist servitude, can only envy and secretly cherish the freer Anglican regime. If it were possible to have two religions, one public, the other private and hidden in the heart, then "most considering Men, whatever their first were, would make ours the second, if they were well acquainted with it."

If mediators help to determine what kind of God they are mediating, then the deity emerging from Sprat's view of the Royal Society's third-party role with the Creator, its replacement of unmediated auto-intoxication, would surely resemble in certain respects an English man of science. At the time, religion colored all aspects of a society simultaneously beset by secularizing forces. And some of those worldly forces were used as vehicles to navigate such a complex sacred-profane culture, to give new insights into the divine. A crisis of mediation opened the door to a portfolio of in-between agents satisfying the human desire to make contact with the superhuman.

Theologians tend to be unfriendly to the idea that people merely "project" their own wishes and yearnings onto phantoms. But they are equally disinclined to say such apparitions exist in the literal sense, if only because beliefs about them can be contradictory. There is a long tradition, however, which suggests that interest in these surrogate deities, and the desire to communicate with them, serves a genuine spiritual purpose, even playing a significant role in the development of religion. In early Hebrew and Christian theology it was not denied that various kinds of spiritual powers are active in the world, though many were regarded as demons, poor specimens in comparison with the supreme God. Christian thinkers in the first stages of their movement believed humans had been around for only a short time. The first man, Adam, had a close rapport with God, but after the Fall his descendants turned away from the original true knowledge to other, false gods. Lacking one-on-one familiarity with the Creator, they needed substitutes, middling beings not of heaven but not of this world either. Keith Ward, onetime Regius Professor of Divinity at Oxford, proposes that the gods of primal religion are imaginative transformations of what people most value. They are "channels of spiritual power," which can change a person's life. Sometimes they operate through dreams and visions. "Spiritual power," as a conveniently vague term, is held to have a real existence, though it is always mixed with the mind's capacity for grand imaginings. And for that reason these strange forces will be met and interpreted in terms of a particular culture.

The "short time" the early Christians calculated had elapsed since Adam and Eve were driven out of the Garden of Eden underwent a drastic expansion with the advancing knowledge of the fossil and archeological record, from thousands to millions of years in the life of mankind. So there must have been immense stretches of existence in which an inkling of the divine had to be won through contact with ill-defined supernatural entities. Even after the great religions had gained a foothold, people believed in the existence of spirits of one sort or another. To call this idolatry is too glib a verdict. Something more interesting is going on. The subgods of this tradition are, in Ward's words, "mediators of the divine, channels and representations of divine power and value." A study of primal religions shows it is quite normal for people to see themselves as part of a sacred system in which they are surrounded by otherworldly powers who represent their longings and ideals. Images and stories, rituals and ceremonies, are a link to these higher presences. The cult of saints, so prevalent in the

Middle Ages, is a later version of this structure. In tribal societies, shamans were mediators of the mediators, enabling members of the tribe to share in the power of gods and goddesses. They entertained revelations from spirits in dreams or trance states. The roaming bands of "prophets" who appear in the Bible, singing and dancing in euphoric rapture and speaking in God-sent utterances, are very similar to the shaman cults.

A "high god," fairly withdrawn and remote, an architect deity who made the world order, has been found in the pantheon of certain tribes. More common, as Ward describes it, is "a riotous plurality of presences— spirits, demons and ancestors, of good and evil intent—whom the shamans or designated mediators can partly control or influence. If a supersensory reality reveals itself to humans, it does so through enhanced or altered conscious states, and in a myriad forms which are not system- atically rationalized."

What does emerge from the study of archaic modes of contact with the divine is that encounters occur within the context of a particular culture, using the media and mental resources of that culture. When gods appear, in whatever form, we cannot say they literally exist. On the other hand, it is too simplistic to say they are mere illusions, a human projection onto a notional deity. These visions of the divine are works of the imagination but "actual" in the sense that, like poetry or music, they transform the ordinary into the extraordinary, making it more real, bringing its inner spirit to the surface.

As in art, a conjuring of the superreal may occur, using many kinds of cultural materials. The history of mediation shows that revelations are not given in the form of explicit, transparent truths guaranteed for the lifetime of the world, though that position was strongly held in the twentieth century by neo-orthodox theologians like Karl Barth. Reve- lations are not always clarifications. We must accept that God did not decide to communicate in plain text; there are his out-of-character pro- nouncements to shocked audiences in the Old Testament. The upshot is that revelation, filtered through whatever vehicle, does not act to dis- solve arguments and perplexity. Quite the opposite. It may leave more arguments and perplexity, though perhaps of a different kind. In the Bible, the prophet Jeremiah (chapter 28) had a quite violent quarrel with another prophet, Hananiah, over the fact that God had spoken differently to each of them. Revelation is not timeless and placeless. It takes place in a specific mental climate at a particular cultural moment. Says Ward: "Instead of thinking of God (assuming for the moment that there is one) as breaking into a human framework, ignoring it completely, and giving direct Divine knowledge, it seems more plausible and more in keeping with the actual history of religions to think of God as communicating within the framework that societies have themselves developed." You might expect, therefore, that God sets about revealing his ultimate divine purpose in terms of the interests and aims of a specific society.

At the time of the early Royal Society, the prevailing culture was a not always consistent mix of the sacred and the secular. There was a half-skeptical, half-credulous attitude toward the supernatural, sometimes both co-existing in the same person. The emerging medium of the novel reflected that dual aspect, and was another vehicle linking the two worlds. In Daniel Defoe's *Robinson Crusoe*, the hero of the novel, thrown onto his own resources on a desert island, is the epitome of man as maker of his own destiny. But he also mirrors the belief of the author in the existence of supernatural beings and their role in the lives of ordinary people. In Crusoe's world, mediators throng about him: good ones steer him to saving paths and bad ones plot to do him harm. Defoe, who was not shy of celebrating the primacy of reason elsewhere, wrote that God does not want for agents; he has apparently "posted an army of ministering spirits, call them angels if you will, or what else you please; I say, posted them round this convex, this globe the earth, to be ready at all events to execute his order and to do his will."

In the seventeenth century there were many documentary accounts of supernatural apparitions, sometimes explained as examples of God's omnipotence, his power to communicate in any way he chooses. These events were regarded as "truer than fiction," mirroring a faith that material things could mediate nonmaterial truths. Apparitions seen and heard, encountered as if in the presence of an actual person receiving a literal revelation, seemed to suit a society caught between an obsession with spiritual matters and a new respect for the evidence of the senses, due to the successes of a science that believed a theory is only as good as experiment demonstrates it to be. The trouble was that the two kinds of knowledge were beginning to separate. Officially, the Royal Society did not believe in ghosts, and the general ascendance of scientific reason made religious or spiritual knowledge and experience rather suspect.

In Thomas Sprat's "history," there is a fierce dislike of what was then known as "Enthusiasm," by which was meant a form of devotion claiming to be in touch with God directly on an open line, no facilitators needed, and dispensing with the assistance of priests or scriptural authorities. The term was intended to be derogatory, denoting an emotional, effusive piety entirely at odds with the cool, cerebral religion that mainstream Anglicanism provided. John Locke, a friend of Newton's, had enthusiasts in mind when he railed against those who "flatter'd themselves with a perswasion of an immediate intercourse with the Deity, and frequent communications from the divine Spirit." Enthusiasm, said Locke, "rises from the conceits of a warmed or overweening brain." Often Enthusiasm represented an eruption of impatience with that same Church, its *sang-froid* and maddening poise, and its disdain for dabblers in the occult. "Enthusiasm" was an umbrella word applied to fringe groups like the Ranters as well as movements as impressive as the Methodists, who refused to compromise with their belief that Christianity was moving in the wrong direction. The attitude is all or nothing: no interpretation, no accommodation, no trimming or getting into bed with secular powers. Ronald Knox, a Catholic chaplain to Oxford University noted for incubating the religious talents of his protégé, Evelyn Waugh, made a special study of Enthusiasm and decided that one of its chief characteristics was a repudiation of mediators in all their manifestations, whether human or quasi-divine. Enthusiasts do not reject the sacraments, but the whole emphasis is on direct personal access to the Author of Everything.

In Enthusiasm, liturgy and theology are minimized. Out goes the idea that human reason is a guide to religious truth. In its place is a "morbid distrust" of the intellect. Art and music, as a means of lifting the mind heavenwards, were thought to impede, not nurture, an immediate relationship with a God available to select souls gifted to communicate with him in this unobstructed manner. The enthusiast believes he or she has come into a "new order of being," with a shiny new set of faculties fitting this exalted state. Enthusiasts may be regarded as a menace to civil order because they tend to regard themselves as citizens of another kingdom, owing loyalty only to it.

A symptom of Enthusiasm is the ecstatic state, its participants "speaking in tongues," breaking out into incomprehensible speech in a language confirmed as unknown by philologists, or slipping anesthetized into a trance without sight or hearing, or quivering with the whole body not for moments but for hours. At John Wesley's revival meetings, members of the audience might burst into songs of praise or fall down in a fit of convulsions, shout out that they were sinners or announce that they were royal kings. This did not sit well with sophisticates of the period. Henry Fielding, in his novel *A History of Tom Jones, a Foundling,* published in 1749, singles out Wesley's partner, George Whitefield, for special mockery. Fielding made clear his own moderate Anglican leanings in the portrait of Squire Allworthy, a man broad-minded and compassionate, his feet planted firmly on common sense, his benign nature evident in his good works and decent opinions. The author is pitting hot religion against cool, and coming down hard on the side of cool. A man could parrot Whitefield's sanctimonious phrases one minute and pick your pocket the next. In *Joseph Andrews,* Parson Adams speaks of Whitefield as having called "Nonsense and Enthusiasm" to his aid.

Ronald Knox, who christens the Enthusiastic movement "ultrasupernaturalism," notes that it became the major preoccupation of religious minds at two junctures of history: in England in the seventeenth century and in the United States in the twentieth. Certainly, in Thomas Sprat's time, there was a strong reaction against enthusiasts. The Quakers were deeply suspect in the early days of their formation. George Fox, their leader, who delayed the foundation of Durham University out of a belief that all learning is vanity and foolishness, decided that churchgoing was superfluous since God was present to every person, a highly subversive attitude. As a twentyish shoemaker's apprentice from Leicestershire, Fox struck out from home in the 1640s and bummed around England, listening to the opinions of ordinary people. He noticed that in some of them there was a sort of inner light, a personal source of truth, a self-tutor, giving knowledge that scripture by itself could not match. Many of the early Quakers were workers on the land, discontented, seething over the steep rents charged by landlords. A note of protest and nonconformity marked the fledgling movement, a new version of an older radical rebellion against Catholic ritual and intermediaries between God and mankind, be they priests or the body of Christ supposedly present in the bread and wine of the Eucharist.

There was something almost shocking about the rapid growth of the Quakers and their robust self-assurance. At the start, Quakers were definitely not pacifists, as they later famously became. In the 1650s they looked threateningly radical and confrontational. At one point General John Lambert was thought to be arming the Quakers in a panicky reaction on the part of a troubled republican regime. Fear of social revolution, egged on by enthusiasts, may have contributed to the sudden shift of public opinion leading to the Restoration of Charles II in 1660, after which the Quakers adopted a pacifist outlook. In the months before the return of the monarchy, fear of the Quakers intensified. At Oxford in July 1659, the sound of a bugle scared the congregation in Carfax Church into thinking "the day of judgment was at hand" and that "quakers were come to cut their throats." When General George Monck, a longtime adversary of the movement, who had purged his army of about forty Quaker soldiers, marched from Scotland to London in 1660, he warned of the peril to society of "fanaticks," a term interchangeable with "enthusiasts." The political opposition to the Quakers helped to usher in a new English religion in which God became more coolly "rational" than ever. In the period following the Restoration there was much less talk about seeing or hearing God in dreams or visions, though it continued among certain fringe groups.

Another factor in the tension between hot and cool, between unmediated and mediated, was the role of women. Women were promoters of Enthusiasm on occasion. They appear in the story of James Nayler, a Quaker leader and charismatic preacher who had served as an officer in Cromwell's New Model Army. In the autumn of 1656, a year in which several predicted the Second Coming would occur, Nayler, recently out of prison, rode into Bristol on a donkey, coiffed and bearded to look like Jesus. A number of women walked knee-deep in mud alongside him, leading his steed by the bridle and crying: "Holy, holy, holy, Lord God of Israel." They kissed his feet and spread their clothes on the muddy road. The ringleader was one Martha Simmonds. Another disciple, Dorcas Erbery, insisted that she had been dead for two days and that Nayler had resurrected her. Nayler was arrested and tried by a committee of the House of Commons. He was sentenced to be branded on the forehead, a hole put in his tongue; also to be locked in a pillory near the Royal Exchange in London. His female entourage, who had continued to attend and kneel before him throughout the trial, grouped themselves around him. One stood behind, two sat at his feet, in imitation of Jesus and the three Marys: Magdalene, the Virgin Mother of Jesus, and Mary the Mother of Cleophas, described in the Gospel of John.

Another religious movement, enthusiastic in nature, known as Quietism, had a strong feminine component. One of its leaders, Miguel de Molinos, was found to have twenty thousand letters from pious women in his possession when he was arrested. He was convicted of heresy and sentenced to life in prison. Quietism was a mystical offshoot of Catholicism. It gave high priority to meditation, with the aim of turning the soul exclusively to the love of God. At the center of Quietism in France was the figure of Jeanne-Marie Guyon, a writer on mystical theology who, as a child, had longed for martyrdom and left her family to wander about southern France for five years with a priest. Married to a man more than twenty years her senior and living with his family, she was tyrannized by her mother-in-law, a lady who unwittingly encouraged her taste for "interior prayer" by watching her from the card table to make sure she did not, seated by the fireside, move her lips. Jeanne-Marie became a widow at the age of twenty-eight, moved to a house of her own, then went to Geneva, taking her daughter but leaving her two sons. From Geneva she toured France and Italy, instructing people in the spiritual exercises of Quietism, partly in the company of a mystical priest who died a lunatic at the asylum at Charenton.

Jeanne-Marie was arrested and put in the Bastille. She was released seven years later and spent the rest of her life near Blois, setting an example of simple piety and plain living. She is celebrated even today as a larger than life figure, the epitome of "hot" religion, even though its name was Quietism. Ronald Knox called the hundred years that followed the Counter-Reformation "an age of introverts," and Quietism was a prime example. It was distrusted by its opponents as a feminine "soft option," apt to hide the fact that being a good Christian is hard work. So hostile was Quietism to mediators that it went a step beyond the Puritan iconoclasts and banned even mental images of the deity. Prayer was an end in itself, a simple looking outward to God.

It was only to be expected that the champions of the Newtonian system, as represented by the Royal Society—almost exclusively male and giving priority to the God of power and majesty over the God of love and caring—should bracket Enthusiasm and the female and deplore the influence of both. Robert Boyle was ostentatiously celibate; Newton avoided sex altogether as far as is known. James I, who insisted on the divine right of kingly power, was a misogynist. The inner light, illumination, mystical communion with the supernatural, prophetic revelation, all were suspect as subverting the progress of true knowledge, and all were female-tainted. A new kind of science religion, which accommodated almost without exception the male gender, went into opposition. It was comfortable with the God of aloofness and mastery. It has been said of these new secular priests—a caste created by the scientific revolution—as the fourth-century father of Church history Eusebius remarked of the rising austere clerics of his day, that they are "above nature, and beyond common human living. Like some celestial beings, they gaze down upon human life, performing the duty of a priesthood to Almighty God for the whole race."

CHAPTER ELEVEN

Simplifying the Divine

"We also are religious and our religion is simple,"
objected the Roman proconsul to the martyr Speratus
at his trial near Carthage on 17 July 180. "If you will
listen calmly," replied Speratus, "I shall tell you the
mystery of simplicity."

—ERIC OSBORN

THE HISTORIAN OF SCIENCE Ian Hacking sprang a surprise on an
audience of distinguished scientists at the 25th Nobel Conference of
1989 in St. Peter, Minnesota, by standing up to deliver a blistering coun-
terblast to monotheism, Newton's axiom of scientific probity.

Recall that Newton not only insisted on the Oneness of the One as
a core theological belief. He also held that monotheism was a salubrious
influence on the study of the physical world. In a cache of papers
entitled "The Original of Religions," he claimed that knowledge of the
creation advanced by leaps and bounds at times when there was a true
idea of God, and stagnated or backslid in periods of pagan idolatry,
Greek metaphysics, hard-core Trinitarianism, and the veneration of
saints. The best conditions for science to forge ahead were those of
primitive monotheism, because only then could it proceed in the confidence that nature was a unity: one God, one creation.

And if the world is a unity, the work of a single master builder, it must at some deep level be simple. "Truth, is ever to be found in simplicity, and not in the multiplicity and confusion of things," Newton wrote. And he added: "It is the perfection of God's works that they are all done with the greatest simplicity. He is the God of order and not of confusion. And therefore they that would understand the frame of the world must endeavour to reduce their knowledge to all possible simplicity." Leibniz, for all his dueling with the Newtonians, was also an apostle of unity, though he saw simple causes producing amazing variety. He deplored the disunity of Europe, especially of Germany, and tried to do something about it. Leibniz looked for consensus: converting China to Christianity not by denying its religion but by pointing out similarities with his own, encouraging scientists to share their knowledge and ideas, inventing a universal language that would obliterate misunderstandings and needless divisions.

In sweeping all this aside, Hacking knew he was speaking heresy. The unity of nature, and science's mission to discover more and deeper unities, is still a sacred principle for many physicists. They sometimes stress the increase in *power* such unity can bring. The eminent physicist John Wheeler, coiner of the term "black hole," has spoken eloquently of "the most basic unifying concept that there is: the faith that the unknown can be made known." That concept made a dramatic leap forward under Newton.

A restless silence greeted Hacking's radical assertion that scientists who adopt "Truth is Simple" as a credo are in fact traitors to the principle that science is a rigorously secular pursuit. They are, he said, wittingly or unwittingly smuggling religious baggage into a discipline that scorns such encumbrances. It is monotheism, Hacking declared, that promotes this pernicious intrusion. He spoke of it much as Francis Bacon had warned against "idols," mental biases and phantoms that distort reality and block the way to truth. "I know of no other reason for thinking simplicity a guide to the truth," Hacking went on. "I suspect that many admirers of unity have, *au fond*, a thoroughly theological motivation, even though they dare not mention God. I wish they would. It would get things out in the open!" As long as people keep alive the myth of the unity of science, he concluded, the accepted story that we have left the theological era behind us is untrue.

At the same conference, the particle physicist Sheldon Glashow took issue with Hacking, but not full-bloodedly or without reservations. Glashow declared his faith that there are simple rules governing the behavior of matter and the evolution of the universe, and that "universal truths" do exist in science. But even he was cautious about affirming the perfect unity and therefore the essential simplicity of nature. He warned that the words "theory" and "truth" are often bandied about too carelessly, even by scientists. Newton's laws of motion and Boyle's law of how gases behave, forerunner of the belief in atoms, are true, but not universally true without exception, Glashow pointed out. They are not monolaws. If you squeeze a gas rather hard, its atoms crowd too close and it turns into a liquid. As for Newton, he did not, could not have known that his laws do not apply in Einsteinian space. So Boyle and Newton, twin gods of the scientific revolution, reign still, but in a parochial kind of way. They are like the pagan deities of antiquity, whose power was not to be underestimated but at the same time was local and confined. Glashow talked about "domains," within which certain laws are correct always and forever. James Clerk Maxwell, who unified magnetism and electricity, wrote equations that hold good today, but whereas he decided that light consists of waves in the electromagnetic ether, quantum theory has superseded that idea and shown that light is neither wave nor particle.

An ideology of simplicity, unifying God and the world, the heavens and the earth, took various forms in the post-Newtonian decades. Newton had written that nature "does nothing in vain when less will serve; for Nature is pleased with simplicity, and affects not the pomp of superfluous causes." It was an invitation to read for "Nature" "God." It sometimes happens that when a doctrine aims to be the ultimate explanation of mankind's predicament, its proponents and champions are gripped by an irresistible urge to simplify. The popularizers of Newton's discoveries, catering to an eager lay audience, succumbed to that temptation. They left out much of the mathematics and concentrated on the Opticks, written in lucid English, rather than the Principia, which was composed in a Latin forbidding enough to discourage dilettantes. After Newton's death in 1727 there was a big market for poems, statues, and simplified versions of his scientific work, including one aimed at gentlewomen entitled Newtonianism for the Ladies which neglected to so much as mention the laws of dynamics. For the ordinary reader all the ungainly features of celestial mechanics had been stripped away, making it easy to admire the compactness and parsimony of his system. Voltaire made his famous comment that Newton had shown nature "nearly naked and made men amorous of her."

Puritan sermons set an example of the spare, uncluttered style. The Royal Society, whose original members were a good two-thirds Puritan, made an explicit commitment to plain language. Thomas Sprat in his "history" wrote scornfully of "this vicious abundance of *Phrase*, this trick of *Metaphors*, this volubility of *Tongue* which makes so great a noise in the world."

The reduced profile of the Trinity in the eighteenth century reflected a popular inclination to get rid of obscure theology. Even though Protestantism was a simpler faith than Roman Catholicism, drastically reducing the number of entities between the individual and the deity, the Puritans had given people more theology than they could stomach. A movement called "Deism" took simplicity to an extreme. The roots of Deism go back to sixteenth-century Poland, where a breakaway sect, the Socinians, published a new catechism, rejecting the Nicene version of the Trinity and insisting on the authority of scripture. In England, during and after the time of Newton, Deism was a "cool" medium of piety, a friend to science and an enemy of Enthusiasm. A diverse collection of "Deist" writers, some more gifted than others, aimed to put Christianity on a neutral basis, purging it of partisan strife and squabbles over small distinctions and minor points of interpretation. It was a "Do Not Disturb" doctrine, prescribing a quiet, uncomplicated life. Lord Herbert of Cherbury, brother of the poet George Herbert, was a chief instigator. He set out a list of "Common Notions" stating that there is one supreme God who ought to be worshipped; that worship is mainly in the form of virtue and piety; that people should be penitent for their sins; and that there are rewards and punishments in this life and the life to come. These truths apply to all human beings at all times and everywhere.

Absent a strong element of the supernatural, the emphasis in Deism came down heavily on practical morals and decent behavior, neither of which was in overabundant supply in Restoration England. Increasingly, "decent behavior" meant conduct that was socially acceptable. Deists felt that the here and now was not to be treated lightly. They thought

God and scripture had provided enough penalties and remuneration in *this* life that they could not be ignored. Promoters of Deism tended to minimize revelation and focus on truths the writers of the Declaration of Independence called "self-evident." These were public and universal truths, as opposed to the private and local disclosures given to God's elect in states of intense spiritual rapture. They were not cloaked in obscurity, surrendering their secrets only to a privileged handful of souls. What has been called "the signal gun of the deistic controversy" was a book, *Christianity Not Mysterious,* by a rakish character named John Toland. It gave considerable annoyance to divines, whose authority rested on the axiom that Christianity is mysterious. It is not much of a stretch to say the book was a herald of the Enlightenment in England, the "Age of Reason."

Toland was an Irishman, a Protestant convert who was chased out of Ireland, his book burned by the public hangman by order of the Irish House of Commons. He wrote like a popular journalist, saying he was educated "from my cradle in the grossest superstition and idolatry." He embarked on an impecunious career as a militant freethinker and, in his own term, a *Cosmopoli*, a man of the world. He left his biographers precious little to work with, leading one to complain that if Christianity is not mysterious, John Toland certainly was. He was a cheerleader for republicanism, but acquired aristocratic patrons. Rumors held that he was a spy for the Prussian court and the paramour of the electress Sophia. He was probably a member of at least one secret society; they were numerous at the time. Toland was obsessed with religion, but personally he was inclined to atheism. He extolled reason, but dabbled in the mystical writings of Giordano Bruno; and he was interested in the Latitudinarian movement, which was extremely unfriendly to him.

Revelation, which mainstream clerics insisted must accompany reason, Toland treated with respect, but in such a way as to cast serious doubts on its supernatural efficacy. He called it a medium of information, which is certainly a less glorious description than that of theologians. He pointed out that in the Bible, leading figures are apt to be wary of accepting revelation at face value, right off the bat. Thus, reason always plays a role, whether we like it or not. The Virgin Mary, "of that Sex that's least Proof against Flattery and Superstition," did not believe she would bear a child of whose kingdom there shall be no end "until the

Angel gave her a satisfactory Answer to the strongest Objection that could be made." Even then she was a little dubious, "unlike her present Worshippers," Toland waspishly threw in.

Deism invented its own divinity. It became an anti-mystery religion, which defined God as an absentee maker of laws that can never be altered. Paying respect to Newton, Deism worked against the idea that a God who creates a clockwork universe, running on fixed laws, adds extra laws in the form of moral decrees, and broadcasts them by revelation. Natural reason, implanted in all at birth, is a sufficient guide to virtue. One writer of the time called this the "Naked Gospel." It was naked of mediators, since it was entirely up to each person to lead an ethical life. Deism wanted an impartial God, one that did not play favorites, as Yahweh sometimes did, and that meant a God of noninterference. Denis Diderot, editor of the famous French *Encyclopédie*, was an enemy, like Newton and Locke, of highfalutin metaphysics and no friend of the doctrine of revelation. He took the Deist side in France and complained that dogmatic Christianity was "unsociable in its morality," implying that it used its own private line to God to separate itself from the rest of humanity. Deism was too pure, too unselfish, too tolerant, and too *un*peculiar to produce evil.

It was almost as if the Deists were saying: Listen. The universe is harmonious, benign, orderly, and in a sense *good*. That is what Newton showed us. So if God is the omnipotent Author of such a universe, he must have created humans along the same lines. In which case they are certainly capable of becoming as good as the universe. Once the distortions of a corrupt tradition have been removed, once the causes of intolerance and doctrinaire bias are gone, this truth will be recognized by all.

Alas, the world does not work that way. Wickedness, it turns out, is as universal as good. Deism's great flaw was that it had a weak sense of the tragic, and no real explanation of the presence of evil in history. Once upon a time, it had been possible to saddle the blame for evil on malicious phantoms, witches, Satan, and other malevolent powers. King James I wrote a treatise on such agencies, called *Daemonologie*, in which he equated them with evil and contrasted them with the divine nature of kingship—though he also exploded more than one fraudulent accusation of witchcraft. But when science helped to clear the air of these superstitions, there was the awkward conundrum of why a monotheistic, omnipotent God allowed evil to persist. And it was a basic proposition of Deism that God *was* omnipotent. The greater the emphasis on God's power, the more intractable the quandary became.

This failure to come to grips with the reality of sin and wrongdoing crushed the illusions of one notable American Deist, Benjamin Franklin. As a young man, Franklin was attracted to the doctrine, but he never could be fully satisfied with just one Supreme Being, portrayed by Deism as impersonal and far off. What really mattered to him was virtue and doing good to one's fellow creatures, which Deism promoted. But Franklin learned the hard way that a theology of this sort, on its own, is not cogent enough to produce virtue in its adherents. Human nature is made in a certain way, with inherent frailties, and Deism seems to underestimate our regrettable inclination to err. Franklin converted two boyhood "friends," John Collins and James Ralph, to the movement. Collins borrowed money from his mentor and scooted off to Barbados without repaying it. Ralph, a dreamy ne'er-do-well who traveled to London with Franklin, also ran up debts that were never settled. Neither seemed troubled by pangs of conscience or fear of reprisals on the part of the Supreme Being. In London, Franklin worked at the celebrated printing house of Samuel Palmer, and assisted in the publication of William Sollaston's *The Religion of Nature Delineated*, which put the case for studying science and nature—rather than blindly accepting divine revelation—as a route to the discovery of religious truth.

"I began to suspect," Franklin wrote, "that this doctrine, though it might be true, was not very useful." Part of the problem was that the God of Deism, so grand and distant, did not seem concerned with the petty affairs of humans and their struggle to be good as the world knows good. Franklin toyed with the idea that to bridge this divide, the Supreme Being caused there to be minor gods, less imposing, but more approachable and easier to worship than he himself. God could manifest himself in a number of different ways, depending on the special requirements of the worshipper. For that, Franklin has been accused of polytheism. The consensus seems to be that he was speaking figuratively but making an important point, that a religion as austere and "scientific" as Deism creates the need for other and sundry mediators.

Franklin was not the only founder of the American republic to flirt with this version of "reasonable" Christianity. Thomas Jefferson, who saw religion chiefly in terms of morals, and George Washington, who wanted no mention of God in the Constitution, were Deists. So, in the vaguer, twentieth century sense of the word, was President Eisenhower.

Arthur Lovejoy, in his masterpiece *The Great Chain of Being*, thought Deist thinkers were "peculiar" in their belief that the most practical truths available to us are simple. To many of them, simplicity "was, in fact, not merely an extrinsic ornament, but almost a necessary attribute of any conception or doctrine which they were willing to accept as true, or even fairly to examine." Toland had called simplicity the "noblest ornament of truth." Sometimes, a sense of the simplicity of what is needed to live a successful and virtuous life went along with a recognition that the universe, more immense and crowded than the medieval one, is actually so complex as to be beyond human understanding. Lovejoy calls this attitude a "pose," but it seems to have been genuine in many cases.

A fashion for the simple spread to other facets of life in the eighteenth century. There grew up a vogue for the primitive, the unsophisticated person whose mind was clear of the artifices of civilization. The phrase "Nature's Simple Plan" was part of the literary jargon of the day. One example of this curious sideshow of the late eighteenth century was the idealizing of the "noble savage." Since religion is universal, and basically simple, so the argument went, and virtue is the birthright of every person, even the untutored savage ought to display Christian excellences. In the autumn of 1774, a South Sea Islander named Omai, or Omaiah, was brought to England by a naval captain. He was presented at court and taken up as a favorite by Lady Sandwich. Omai took a fancy to London society, and society liked him back on account of his good nature and naïveté. He was courteous and his manners were acceptable. He was invited to dinner at the house of Mrs. Thrale in Streatham and introduced to Dr. Johnson, who found the "savage" as polite as any London socialite. Johnson told Boswell that one night at dinner, Omai and Lord Mulgrave were sitting opposite him with their backs to the light, so he could not see them distinctly. "Sir," Johnson said, "there was so little of the savage in Omai, that I was afraid to speak to either, lest I should mistake one for the other." Omai even beat Baretti at chess and behaved better in victory than Baretti did in defeat. He was presented at court and given an allowance by George III, whom he addressed with impeccable deference as "King Tosh." Fanny Burney considered that Omai's gracious manner "shamed education."

Alas, the finale was anticlimax. When the time came for Omai to go home, there was a sense of double letdown. It was said he had not acquired the sort of cultivation that went deep and was likely to last; on the other hand, he was ruined for life in his native habitat. "It was not a question of measuring civilization by its material persuasions," wrote Chauncey Tinker, author of a study of the Omai affair, "but rather of testing men by their capacity to make a proper use of such 'blessings.' In the last analysis, it is the capacity which distinguishes an Omai from a Pericles." The organizers of Omai's visit had been suffering from an overdose of optimism.

Another anticlimax of Enlightenment thinking was the virtual collapse of a dream: that of creating a "social physics," a science that would reduce the complications and ambiguities of human society to the kind of basic calculus that had proved so powerful in the case of the laws of dynamics. Grand conclusions could supposedly be deduced from a handful of "naked" axioms about human nature. It was to be as compact and logical as geometry. One of these axioms was that society is essentially simple. Locke, a dedicated Newtonian who grew up in a liberal Puritan home where simplicity was highly prized, developed this premise, neglecting to appreciate that people are not simple at all. It was thought that Locke was better at overturning long-held beliefs than at constructing new ideas to put in their place. The social sciences of the eighteenth century became somewhat sterile, much as theology had after the Restoration, unable to rise to the task of dealing with the shifts and dislocations of the industrial age.

A corollary of such "geometrical" thinking was that knowledge, being essentially simple, is available to all and sundry and therefore should be an underpinning of an egalitarian society. A simple science is a democratic science. It is also to some extent *natural*, since it appeals to our ordinary understanding. That was the view of the marquis de Condorcet, one of the intellectual Fathers of the Idea of Progress, the doctrine that the world is improvable. Francis Bacon had been instrumental in fostering this notion with his belief that we must turn our faces from the past toward the future. Some of Condorcet's contemporaries were not so sure. A suspicion began to emerge that he was profoundly wrong.

Science is an *unnatural* activity, speaking the artificial language of mathematics. Scientists talk about "a principle of simplicity": when more than one possible law is derivable from the same set of observations, the principle says select the least intricate one. But there may be an array of possible laws, each roughly as simple as the others. In that case a scientist decides on the basis of a theory, which may not select the simplest option.

Einstein defined his deity as a God of Parsimony and gave the impression that this God only lights up our understanding when we have made a strenuous effort to render our ideas about the universe as simple as possible. But while the logical underpinnings of Einstein's law of gravitation are simpler than those of Newton, the formulation itself is awesomely complex. Einstein uses fourteen equations as against Newton's three. Such deep complexity is masked, however, by the beautifully neat surface concept that gravitation and inertia are equivalent.

Paul Dirac, the brilliant English theoretical physicist who is often called the "twentieth-century Newton," made it his basic creed that an inherent harmony exists between the austere plainness of mathematics and the operations of nature, which abhors surplus paraphernalia. Helge Kragh, Dirac's biographer, thinks this was an "aristocratic" outlook—recall that Deism was in part an aristocratic movement—and also a religious one. He uses the word "priestly" in describing Dirac's personality. Dirac himself, whose upbringing had been unusually austere and solitary, suggested that his belief in mathematical beauty was in some sense akin to religious faith. The great physicist Niels Bohr said: "Of all physicists, Dirac has the purest soul."

In his work at Cambridge University, Dirac explicitly opted for theories that are either simple or beautiful, or both. Although Einstein's theory of gravity is much less simple than Newton's, Dirac considered it to be far more beautiful and thus more "true." But his insistence on these two aesthetic principles led him into a trap. When he was looking for an alternative to quantum electrodynamics, which can only be expressed in a highly complicated mathematical scheme, the results were a disappointing contrast to his earlier triumphs in the mid-1930s. In fact, the moment when he emphasized most strongly his belief in the simplicity and beauty of truth was also the moment when his productivity as a physicist began to decline. "It is not irrelevant to point out," says

Kragh, "that the principle of mathematical beauty governed Dirac's thinking only during the later period."

We cannot avoid the suspicion that yet another anticlimax had beset belief in the simplicity of nature. And that if such a belief in this instance was quasi-religious, the idea of physics as a "calling," a mediator of the divine, might be joining other formerly sacred concepts on the road to full secularization.

CHAPTER TWELVE

"People Simply Cannot Be Religious Any More"

My dear friend, let us look into providences; surely they
mean something?

—OLIVER CROMWELL

Cromwell was about to ravage all of Christendom, the
royal family was lost, and his own set to be ever-
powerful, but for a little grain of sand which lodged in
his bladder. Even Rome was about to tremble beneath
him. But once this little piece of gravel was there, he
died, his family fell into disgrace, peace reigned, and the
King was restored.

—BLAISE PASCAL

THE IDEA, PROMINENT IN the euphoric climate of the Enlightenment, that knowledge can be democratized had the paradoxical effect of
making the secular somehow a little sacred while giving the sacred multiple links to the secular. It was a two-way street, and in one form or
another has remained so to this day. Physics has never entirely lost its
tenuous links to the divine.

In the post-Newton period, a vogue developed for the "ordinary," a
marked shift toward matters of fact and the commonplaces of everyday

life, turning away from the exotic and strange. George Herbert's poems, presenting homely tasks as a form of worship, were an early example of this trend. An embrace of the mundane occurred in religious as well as in worldly affairs. It suited a new mood of conservatism, which escalated into a panic fear of change after the French Revolution, whose potential for social chaos, it was feared, could cross the English Channel and infect the land which had undergone decades of trauma to set its politics on a stable basis.

John Locke had made religion seem quite ordinary. Locke believed in God, but he justified that belief in a cool, intellectual fashion. He accepted the fact of revelation and its significance, but in an oddly backhanded way. Instead of recognizing the mystery and otherworldliness of revelation, its "beyond" character, he recruited it for his opinion that the truths of religion are plain and simple, a matter of common sense. What could be simpler and more lucid than the idea that Jesus came to save the world?

The lure of facts as opposed to speculation and apocalyptic dreamwork tended to promote a partiality for things as they are rather than as visionaries wished them to be. People took a greater interest in the past and in history. With the ascendancy of fact, there was a readiness to accept that some religious truths are only probable. That, too, was an argument against the cocksureness of the left-wing sectarian movements, with their unshakable confidence in the certainty of their own inner voices.

Historians have mostly neglected this elevation of daily life to greater significance, but others recognize it as an important marker of attitudes and values. One sign of the new prominence of the ordinary was the role of lawyers and their special impact on a society in flux. William Bouwsma has tracked the rise of the legal profession in early modern Europe. He found that lawyers ceased to believe, by and large, that systems of justice reflect the will of God, or that laws mediate between sublime wisdom and the particular requirements of ordinary people. Instead, they regarded the law as an institution that responds to everyday humdrum human needs. And lawyers were instrumental in the shift to a secular world. They saw society as complex, not simple, as the "social physicists" and the Deists tended to imagine. And they were worldly enough to recognize that life is a matter of multiple forces in unceasing conflict: it is often better to let that conflict play itself out than to stifle it. "Lawyers," Bouwsma writes, "represented the growing assumption that life in the world is only tolerable when it is conceived as a secular affair, and that the world's activities must be conducted according to manageable principles of their own rather than in subordination to some larger definition of the ultimate purpose of existence." By applying this assumption to constantly changing problems of their societies, lawyers were "the supreme secularizers of their world."

Lawyers also reflected the conservative reflexes that operated in Europe in those decades. The fact that most of them were religious, some deeply so and leaning to the "Heavenly City" ideas of St. Augustine, actually intensified their secularizing effect rather than weakening it. The world could not duplicate the divine order of that celestial city, but the law could go a long way toward creating a better order than had hitherto existed. With order comes a sort of meaning. In a time of tremendous instability and agitation, it reduced the uncertainty of life rather as unquestioning faith had done when religion was at its high tide. Law set limits to a social system that was fraying at the edges. Bouwsma points out that lawyers, being for the most part well paid and conservative, wanted to make society work. That was an eminently practical, this-world approach. "By defining what was socially intolerable and by reshaping the official forms of social intercourse," he says, "laws and the men who worked with them must have gradually renewed that sense of limit in the social universe, so profoundly threatened by the crumbling of established conventions, without which life had become not only practically hazardous, but in a deeper sense unsatisfactory."

Alongside these secularizing currents, there was an inclination in the misleadingly named Age of Reason to think in terms of method and system, whatever the topic in question might be. Where this was the case, religion was treated as if it were something to be taken apart and analyzed, a "thing" about which cool, cerebral theories could be concocted. Hugo de Groot, better known as Grotius, a Dutch jurist, was notable for his calmly reasonable view of Christianity as essentially made up of statements of fact, a set of precepts. Grotius derived some of his ideas from the Stoic theory of natural law. That theory, dominant for a time in the Roman Empire, held that a good fit exists between the reason that rules the universe and the reason humans are born possessing.

Since universal reason is at work in all parts of nature, there must be a criterion of morality above and beyond the specific local decrees of a given society. The limits of a regulated social order were the limits of the cosmos itself: it was a *cosmopolis*, and people who lived in it were "cosmopolitan."

During the early modern period, an increasingly abstract theology, putting God even more firmly at a distance, was opposed by less "intellectual" movements such as Pietism and Methodism. Theories of the universe as a machine went along with a concept of "system," an organized body of knowledge into which everything must fit, and whose completeness was a warrant of its truth. In religion, that meant paying close attention to its doctrines and creeds. Are they true or false? On the answer to that question hung the validity of the religion itself. The American scholar Wilfred Cantwell Smith sees this shift in outlook as taking on a life of its own. "In pamphlet after pamphlet, treatise after treatise, decade after decade, the notion was driven home that a religion is something that one believes or does not believe, something whose propositions are true or not true, something whose *locus* is the realm of the intelligible, and which is up for inspection before the speculative mind." This attitude, Smith says, had by the middle of the eighteenth century sunk deep into the European consciousness. And it was not confined to theology. It was part of a worldview that was cerebral and systematic.

At the same time a hankering after simplicity had given rise to a certain literal-mindedness. Luther thought the words of the Bible retained some of the properties of the original language of Adam. In a sense, he replaced the authority of the pope with that of the Bible, and for the next two centuries its rule was absolute. Unlike Calvin, however, he was magnificently disorganized and famously unsystematic in his thinking, throwing up new ideas, smashing old ones, boldly going where no churchman had dared to go. He was "a mountain torrent, plunging grandly down from the heights in springtime, loosening rubble and mud in its course, destroying, purifying, refreshing." By contrast, Calvin was methodical and highly disciplined, at a time when discipline was badly needed. He studied law for five years at the schools of Bourges and Orléans and was licensed to practice. Calvin introduced order, system, and authority into a faith that was reformed but not yet given a tight structure.

What was happening in the eighteenth century was that Church leaders and theologians were splitting the old idea of "belief" in two. On the one hand was the confidence that God is real, that he exists, powerfully endorsed by Locke. On the other was a trust in a different dimension of God, one showing him as a loving, caring presence, who listened to prayer. The first was increasingly a being of intellect, logic, and reason, which fitted the new sophistication of a cultured public. The historian James Turner argues that a need to bring religion up to date in a fast-changing world marked the difference between the two sorts of piety. The two came apart subtly and gradually, underneath the radar of history. The breakup was not at first clearly delineated but "belief" and "faith," once interchangeable terms, were in practice used more and more in different ways. One expressed an intellectual grasp of "matters of fact," the other a personal trust.

The slowness of the process and the lack of words to describe it meant that the average person could go his or her way innocently unaware that a seismic convulsion was occurring on the landscape of spirituality. The fact that now one could speak of "religions" in the plural, rather than "religion" as a relationship with the divine, made it clear that something had changed. One religion differed from another in terms of the articles of its creed, which could be dissected and studied by a detached observer, weighing its truth or falsehood. Islam might be coolly contrasted with Judaism, or Buddhism with Zoroastrianism. In the Middle Ages, "faith" was the important thing; "religion" was used in connection with specific monastic communities. Wilfred Smith notes that if religion is an impulse of the heart, it ought not to be divisible. "Piety," "obedience," "reverence"—none of these words has a plural. They are exclusively singular. The plural—as was commonly used in post-Newtonian Europe—is possible only when a religion is seen from the outside, when it is converted into a "something," an abstract object that can be dissected and analyzed. In a provocative summary, Smith says he has come to feel that in some ways it is easier to be religious without the concept of "religion," that the very term is an enemy of piety. "One might almost say that the concern of the religious man is with God; the concern of the observer is with religion."

Such uneasiness about a word cursed with a plural form never really evaporated, even in our own time. Dietrich Bonhoeffer, executed by the

Nazis in the closing days of World War II for his part in the plot to assassinate Hitler, became severely disenchanted with the whole concept of religion as it was understood and practiced, not only in Germany but in Christian Europe as a whole. "We are moving toward a completely religionless time: people as they are now simply cannot be religious any more," he wrote to his friend Eberhard Bethge from Tegel prison on December 5, 1943. Bonhoeffer was adamantly opposed to the idea that the Gospels are there to soothe people's anxiety. Rather, they are a highly disturbing and uncomfortable fact. Even in a "religious" country like the United States, where the Church's loss of political and social power is an opportunity to seek a new kind of freedom, today's theologians commend Bonhoeffer's vehemently stated recoil from conventional attitudes. In 1942 he wrote, again to Bethge: "My resistance against everything 'religious' grows. Often it amounts to an instinctive revulsion, which is certainly not good. I am not religious by nature, But I have to think continually of God and Christ; authenticity, life, freedom and compassion mean a great deal to me. It is just their religious manifestations which are so unattractive."

Bonhoeffer assured Bethge that if he came out of prison, he would not emerge as a *Homo religioso*. "My suspicion and horror of religiosity are greater than ever." He was aware that the Church's interest in "keeping up" with new ideas and movements in the secular world could lead it into blind alleys, or even make it look silly. His view was that religious people are often drawn into the fool's game of invoking God when they bump up against the limits of ordinary knowledge about science or life. They use him as a *deus ex machina* when human wisdom is unable to proceed further. Today, we are inclined to push back or eliminate boundary conditions of existence such as sin and death. Bonhoeffer thought people's "religious" sensibility had coarsened to the point where these two flaming guardians of the gates, sin and death, no longer counted for much. "I have become doubtful," he wrote, "of talking about any human boundaries (even death) which people now hardly fear; and is sin, which they now hardly understand, still a genuine boundary today?"

Another theologian at odds with the Nazi regime, whose ideas resonated with Bonhoeffer, was Karl Barth. Both men were strongly opposed to the idea that there is "religion" first and then faith. No one is born with an a priori knowledge of God. We can reason about our faith only after the fact and we can say things about God only in the light of what he has chosen to reveal to us.

Barth's theory is flat contrary to the notion that religion produces God, not God religion. Such an idea came to the fore as modernity got underway in earnest, during the eighteenth century. It was a stark contrast to the medieval sensibility, which treated God as an indescribable presence saturating the whole of existence. Bumped up by the stunning successes of human intellect and the reputation of scientist-theologians, intellectuals grew more poised about speculating on what sort of God had produced our sort of world: the heightened sense of confidence in human powers of knowing, and the new standards of rigor introduced by scientific methods, made such speculations look more compelling. There was a fading emphasis on revelation and grace and less interest in the doctrine of the Trinity, and this worked to make God seem simpler, and more singular, than before, which of course made him easier to describe. A new kind of theology had given birth to a new kind of divinity. A mental image of God had replaced the old painted and sculpted ones smashed and burned by the iconoclasts. "Instead of discussing the character of God's saving activity in human life," says Bernard Cooke of the mid-eighteenth-century religious consciousness, "much of the educated populace of Europe and America began to argue about the effect of *thinking* about a saving God."

Clever minds in the Enlightenment Newtonified the deity, which meant he came to bear a close resemblance to Newtonian intellectuals. "God said, *Let Newton be!*" sang the poet Alexander Pope, as if God had reconsidered his ancient ban on humans acquiring godlike knowledge; as if he had specially created a mortal mind as wise as his. The trouble is that sooner or later that sort of deification of the brainy elite is liable to fall out of fashion, and when it does, God becomes more unlike us than ever.

Barth went violently against any and all attempts by well-meaning Church leaders to fit God's revelation into a preexisting framework. He marshaled all his considerable polemical gifts to discredit the idea—another "natural" stage in the transition from the sacred to the secular—that theology is just an invention of human culture. Harking back to pre-Enlightenment thinking, Barth said that not only our will but our intellect was impaired by the orchard theft of Adam and Eve, so it is useless trying to know God through our own mental exertions. Revelation comes from God, in his own time and by its own choosing. We do not grasp God, mentally or emotionally; rather, he grasps us. That was a huge departure from the Romantic and poetic idea, prevalent in the nineteenth century, that religion is a matter of immediate feeling, an essentially private inward disposition, "pious exaltations of the mind," which are associated with imagination and spontaneity, and a sense of the mystery of things. The poet Samuel Taylor Coleridge had set the scene for this kind of theology, freeing it from the cool, rational spirit of the Enlightenment and basing it in each person's experience of God.

That was not at all the way Barth saw it. The immense successes of science are to be admired, but they have nothing to do with theology; they cannot add to our knowledge of God. They merely clog the mind with preconceptions. If, said Barth, speaking of God's revelation, "we have not trodden on the toes of every single human method of investigation and grievously annoyed it, we have spoken of something else." Religion itself—and here Barth had a strong influence on Bonhoeffer— is subject to divine judgment and is a human, not a purely divine thing. Religion can be arrogant, trying to master the world and make its adherents feel secure. That is a pipe dream, because God is not to be captured in any human net. Far from being a deliverance, religion is often a self-constructed prison "precisely where human beings bolt and bar themselves against God."

In fact, World War I persuaded Barth that the God religious people worship may turn out to be nothing more than an idol. When the war broke out, he was shocked to see his teacher, Adolph Harnack, ennobled by the Kaiser, organize a petition of nearly a hundred German intellectuals to tell the Kaiser they were firmly in support of the war effort. A preacher in Berlin told his parishioners that Germany was going into battle "for our culture against the uncultured, for German civilisation against barbarism, for the free German personality bound to God against the instincts of the undisciplined masses." Barth thought this ripped aside the mask of Romantic, liberal theology linked to the culture of the times, blatantly exposing the diseased body of "religion." It led him to reassess all his ideas about Christian ethics and doctrine. Early in 1916, he gave a talk in the town church of Aarau in which he called the war an "atrocity" that made people reach out "like drowning men grasping at straws," for the certainty that conscience is supposed to give, but which is dulled precisely by our religious exertions. Conscience is a mediator; but like many of its kind, it may be clouded over by its quasi-human character. Thus conscience turns into an idol as well. The God we think we know "cannot stop those who believe in him—all those excellent American and European representatives of culture, welfare and progress, all those upright, zealous citizens and pious Christians from falling on one another with fire and murder to the amazement and derision of the poor heathen in India and Africa."

To Harnack's statement that God must be understood on the basis of culture and ethics to protect these tenets against atheism, Barth, in a fury, replied that such ideas "come out of polytheism." When the Nazi Party came to power in 1933, Barth was again driven to ferocity by the tepid or worse than tepid reaction of the German churches, and repeated his belief that "religion" is liable to confer its prestige on idolatry and falsehood. A glowing endorsement by a committee of three leading churchmen a few weeks after Hitler was elected Reichschancellor announced that the new regime had been given to the people by God.

Barth saw this as a violation of his deepest principle, that the Word of God must never, ever, be wedded to an ideology. It must never appeal to culture or human nature or politics to justify its truths. The fellowship of the Church, he wrote in a pamphlet sent to Hitler, is not determined by blood, or race, but by "the Holy Spirit and Baptism." He was soon in trouble with the regime. He refused to comply with an order that all university lectures must begin with a "Heil Hitler!" salute. Instead, Barth began his with a prayer.

In 1935, Barth left Germany for Switzerland. Though he changed his mind on such matters as baptism, and could be wayward, he never deviated from his core belief that theology must under no circumstances be a captive to culture or compromise with principalities and powers. And he stuck to his early insight that the message of the Gospels has nothing to do with our finding God out, but instead is the Word coming to us as a verdict on our own pretensions, "including our religious pretensions." God does not say yes. He says no. Human nature is not continuous with the divine but utterly different, totally other. The distance from earth to heaven is immense. That means our religiousness can never carry us across such a vast abyss. Religion is no superjet with unlimited range; it can never give us the buoyancy of faith, and neither can our claims to goodness. Faith comes as a stranger, from another realm, and as a surprise. "The Bible tells us not how we should talk with God but what he says to us."

CHAPTER THIRTEEN

Secularizing the Sacred

We adore the mysteries of the Godhead. That is better
than to investigate them.
—PHILIP MELANCHTHON

A MONG THE RELIGIOUS THEMES that moved into the secular
realm of the "ordinary," three offer strong evidence that domestication
can tame the most exotic and otherworldly expressions of divine will.
These three are: God's providence, predestination, and the millennium,
all flourishing articles of belief existing side by side with the revolution
in knowledge at the birthtime of modernity.

Providence was vividly real, for example, to Oliver Cromwell. It was a
factor in his decision to cut off the head of Charles I. And an urgent question was whether providence chastised wrongdoers in this world as well as
settling scores in the hereafter. A contribution to this debate had been made
by Dr. Thomas Beard, the master of Cromwell's school at Huntingdon,
near Cambridge, in a book, *The Theatre of God's Judgements* (1597), published
two years before Cromwell was born. The word "Providence" originally
implied that a supernatural power or powers "provide for" humankind,
with knowledge of and concern for their future. In Beard's treatise, however, the emphasis was on penalties visited in the here and now on people
who violate the moral code. And that includes kings, who are apt to consider
themselves above the law and beyond the reach of retribution.

Providence is a recurrent theme in the career of Cromwell. As a military commander in the English Civil War, he often attributed success in battle directly to its mysterious workings. As early as 1643, he led an attack against superior Royalist forces at Grantham in Lincolnshire (Isaac Newton, then one year old, was living nearby at Woolsthorpe), and having routed his opponents stated in his report of the battle: "God hath given us, this evening, a glorious victory over our enemies." The fact that he was able to prevail, though outnumbered by enemy forces, reinforced Cromwell's belief that too much attention is paid to "men and visible helps" and not enough to God's guiding hand. After the Battle of Marston Moor, he declared of the Royalist forces: "God made them as stubble to our swords." Of great significance for history is the fact that Cromwell was firmly of the opinion that if things turn out well in any risky enterprise, it is a certain sign that it was crowned with God's blessing and was part of his providential plan. Sometimes he would play a waiting game, holding still until it was clear what God's intentions were to be, perhaps in the hope that providence would decide for itself situations fraught with hazard for the political leader of a state in turmoil.

One of the matters on which Cromwell was deeply undecided was the execution of the king. Certain clerics were taking to the pulpit to quote texts from the Bible supporting regicide, inflaming popular opinion. Their message was that executing a monarch who had wronged his people was nothing more than executing God's providence, his correction of sinners in this world. But Cromwell himself was in no rush to endorse such fanatical sentiments. He wanted a trial of the king, and made plans to talk with him. Historians believe he was genuine in his reluctance to take the extreme step of a beheading. At the very least, he was undecided. He tried to calm down the radical firebrands in the Army Council. But in typical Cromwellian style, when it became clear that political forces were sweeping opinion with unstoppable momentum in the direction of killing the king, he made an abrupt U-turn, using the theme of providence to justify what was actually a stunning shift of position. He announced in Parliament that "Since the Providence of God hath cast this upon us, I cannot but submit to Providence, though I am not yet provided to give you my advice."

When Charles was at last beheaded at Whitehall (wearing two shirts so he would not shiver in the chilly weather letting spectators think he was trembling with fear), Cromwell, it was said, was at prayer, giving the impression he was wavering a little, waiting for God's final say, like a prison warden on hold in case of a last moment stay of execution. Yet he had already signed the warrant of execution, with every sign of impatience with members of Parliament who wanted a treaty with the king. He quoted from the Book of Isaiah, speaking of the thunder of God's wrath on those who thwarted his plans. At the news of the king's death, according to one report, Cromwell held up his hands and declared it had evidently not been "the pleasure of God that he would live," as if providence had intervened on its own account.

Cromwell's God was highly interventionist, always active, keeping his own life and that of England on the right track. Cromwell was apt to lapse into a fury when it was suggested he was inventing spurious pressures to act in a certain way, and then describing them as forces of God's overriding providence. A Royalist poet, Joseph Beaumont, clearly with Cromwell in mind, wrote of a brutish general who "Prints on his mad adventure's exigence/The specious title of Necessity." In a rage, Cromwell felt the need to tell Parliament that "feigned necessities, imaginary necessities, are the greatest cozenage that men can put upon the providence of God, and make pretences to break down rules by."

What was happening was that the concept of "providence," essentially a religious idea, was losing its otherworldly character, coming down to earth as a suspect doctrine that could be used for specious purposes and devious stratagems. Cromwell's exposed nerve in this matter, the violence of his denial that he was insincere, shows up the shakiness of a once potent sacramental theme.

Providence, first cousin to predestination, had been a doctrine prized by the Church and also by superstars of the scientific revolution, Robert Boyle, John Wilkins, Isaac Barrow, all saw purpose and accord in the universe only because the providential hand of God was operating in it at all times. It was a "caring" universe, and therefore profoundly unnatural. Calvin, as usual celebrating God's energy, his ceaseless activity, made divine providence look all the more formidable by stressing its sometimes unreadable character. Providence is part of a *secret* plan, which is always in place, evidences to the contrary notwithstanding. What appears to be chance or accident is really divine providence, whose expression baffles our slow and simple minds. The order, method, and purpose of such events are "hidden in the counsel of God." At his most vituperative, Calvin scorned the doubters of providence, saying that they "will learn, when it is too late, how much better it had been reverently to regard the secret counsels of God, than to belch forth blasphemies which pollute the face of heaven."

What if providence is sometimes baffling to human intelligence? Calvin had no high opinion of that human faculty. In his magnum opus, the *Institutes of the Christian Religion,* he called it "sluggish and groveling." Left to itself, it would turn the world into a chaos of warfare and destruction. Wars happen, even under the umbrella of providence, but they are decided by God, which is why inferior troops sometimes defeat better-equipped ones. That is a lot for God to do, but Calvin repeatedly stressed the work ethic in his theology: it would be disgraceful for God to sit in heaven at leisure as he did on the seventh day of Creation. It would insult the doctrine of his omnipotence. It would make him look womanish. Providence is action, and God is a God of action.

As an idea, providence carried the religious sense of a plan so awesomely grand the finite minds of humans could discern only bits and pieces of it. Without the concept, no coherent sense at all could be made of world history over long timespans. That was grasped by one of Louis XIV's religious brainstrusters, the erudite Catholic bishop Jacques-Bénigne Bossuet, a champion of the divine right of kings, the doctrine which stated that royal authority is so sacred that any attempt to overthrow it is a crime. In his *Discourse on Universal History,* Bossuet recruited the theme of providence to connect up the apparently disjointed moments and episodes of history, binding them into a continuous running narrative, in which there is a clear improvement of the human condition as past gives way to future. The historian Robert Nisbet in a masterly commentary noted that while Bossuet insists that the basis of human advance is religion, he finishes up the book in a decidedly worldly vein. Surveying the rise and fall of vast empires—Scythians, Ethiopians, Egyptians, Persians, Greeks, Romans—Bossuet does not explain their careers solely in terms of God's mysterious plan. Instead, he points as causes of rise and decline to secular factors like economics, culture, and politics. Providence is still intact; but Bossuet is interested in the secondary causes by which it actually works in the world.

This was a landmark event. The temptation, from the seventeenth century on, was to secularize providence, in however hedged and cautious a manner. Rather than speculate on how God operates directly in his saving schemes, sophisticates of the period preferred to talk about what happens when people think about the *idea* of a providential deity, a distancing move if ever there was one. Somehow, things go *as if* divinely guided. That was a principle similar to the economic theory of Adam Smith, that an "invisible hand" causes the superficially untidy and selfish decisions of individuals to lead to beneficial results overall. The same idea is implicit in the American Constitution.

Science was a key component in this move to domesticate divine providence into something more worldly. In the extreme, there was a sense that it could more or less be replaced by the laws of physics. The philosopher Baruch Spinoza, expelled from the Amsterdam synagogue for freethinking and atheism, was an early advocate of this quite radical idea. Working by day at the menial and unhealthy occupation of lens grinding, Spinoza redefined providence as "nothing but the striving we find in both Nature as a whole and in particular things, tending to maintain and preserve their being." "Nothing but" is a dangerous form of words, and Spinoza used it repeatedly. God's providence is nothing but nature's rules. It would be stretching a point to say people in general accepted the idea that providence is just another term for laws of nature, but the rise of science and its willingness to poach into theology, not to mention the readiness of theology to return the compliment, fed such heresy.

Robert Boyle, who endowed a series of lectures promoting "reasonable" Christianity slanted toward moderate Anglicans, was so disturbed by this naturalizing of the otherworldly that in about 1675 he penned some essays on miracles and divine providence. But these sincere men were performing a wobbly tightrope act. The dividing line between God's plan and the operations of nature began to grow fuzzy. Boyle and his like had the difficult task of marrying the two without veering off into profanation, as Spinoza seemed to be doing. Newton's physics won him acclaim in part because it was based on a faith in the essential simplicity and regularity of nature, and also embraced the great simplifying concept of providence. The God of science was also the God of history.

The doctrine of providence came in useful as a justification for quite high-handed human manipulation of history during the so-called Glorious Revolution of 1689, a bloodless, gradualist, but wholly amoral coup which dislodged the reigning monarch, James II, a Roman Catholic, and installed in his place William of Orange and his wife Mary, the eldest daughter of James. A ruling-class clique assisted in William's decision to invade the country; the Lords and Commons offered the two of them the throne as long as they agreed to observe certain constitutional codes. As it turned out, Parliament was given a decisive place in the new government. The affair was managed with what one historian has called "cynical audacity." Throwing out a legitimate king and justifying it by an appeal to "the sovereignty of popular will" introduced large helpings of ethical ambiguity. How could such a suspect deposition be reconciled with God's plan? One answer, an oddly lame one for what was supposedly an Age of Reason, was that providence had marvelously come to the rescue by filling the vacancy left by the fleeing James with the duo team of William and Mary.

Newton was in the front lines of the opposition to James II, and welcoming to the Glorious Revolution. The Newton scholar Margaret Jacob makes an ingenious case that his theory of a secular-religious universe in harmony with itself and under the supervision of an active deity helped to explain away the inglorious ethical questions involved in switching heads of state in this way. If the dubious shenanigans of shopping for a new king could be seen in the context of the logic, order, and reason of the Newtonian cosmos, where an all-powerful God worked his mysterious but ultimately benevolent will, it could be made less distasteful to fastidious souls troubled by the ethics of the whole affair. John Locke had formulated a theory of the social contract, which obligated rulers to think first and foremost of the well-being of the people. If the state blunders into bad decisions, putting the safety of the people at risk, then it is permissible for them to move the state in a different direction—by revolution if necessary. But that did not sit well with churchmen who objected that the sovereignty of the people, carried to such an extreme, was an insult to God's omnipotence, to his sovereignty. The answer was to acknowledge the ultimately divine character of providence, but also to play up its worldly operations, balancing its occasionally amoral, ugly side against the elegant handiwork of the physical Creation. Such a Creator must have an ultimate plan for human history grand enough to justify such blemishes. "The God of order, governor of the universe," Jacob contends, "became the cornerstone of the church's political and social teaching. God's providence operated in every aspect of reality—in the natural order and in the world of human affairs men observed the preserving providence of God."

The doctrine of divine providence was evolving into something much more modern, much more secular: the Idea of Progress. In a classic telling of this story, Robert Nisbet finds that one factor was a newly concentrated focus, at the time of the scientific revolution, on human needs. In a period still saturated with religion, people became "social minded." Radical proposals were made to improve education, purging the universities of old-fashioned and useless scholarship. The Puritans wanted only secular studies at Oxford and Cambridge and extolled the virtues of science, especially in the form of experiment. The humanities they only tolerated at best, urging that greater attention be given to mathematics, chemistry, and geography, and less to the classical texts. Progress, not spiritual but practical, and leading to the improvement of human life, was what they agitated for.

Providence was now a progressive force. And that altered the portrait of the God once supposed to be its sole Author. As the world hurtled onward toward full modernity, God was recruited as an executor of human social improvement. "God," in Nisbet's words, "is no longer seen as a remote, separate, directing omnipotence. He becomes understandable, rather, as a kind of *process*." Not by everyone, certainly, but by numbers of the sophisticated. A "natural and inexorable" pattern of progress began to be discerned. The idea of "grace," special to Christianity, drifted into the notion of "stages of advancement." How easy, then, to just let God "slip away entirely."

One example of the transitional nature of this providence-progress development was the odd emergence of a militant force at the time of Cromwell known as the Fifth Monarchy Men. This group linked a belief in divine providence to a theory of a progressive series of four "monarchies" in world history, starting with the Babylonian and proceeding to the Roman. The Fifth Monarchy came in with the Puritan Revolution in England and would end with the reign of Christ over a world first to be cleansed and made entirely new in a frenzy of final violence; no half-measures in the form of "improvement," but a clean sweep and utter destruction of existing government and law. Its adherents believed this would come to pass according to God's providential design as foretold in the Book of Daniel. On the surface, the movement seemed to be wholly concentrated on the spiritual, on the work of Christ quite independent of human kings or magistrates, and on the role of genuine living saints in a reconstructed world. But Nisbet detects definite secular threads in the fabric of the Fifth Monarchy agenda. Attracting mainly working-class members from the towns, as well as officers, some quite senior, from the New Model Army, it was permeated with anti-establishment sentiment and tunred out sermons and pamphlets that looked to social reforms and the removal of injustices, especially in the matter of arrest and imprisonment without cause. There was a hatred of taxes and tithes, as well as of rents and landlords.

"The ease with which Christ could be supplemented with purely earthly rulers and reformers in the cause of instituting the fifth and golden age on earth is illustrated by the early liaison effected between Fifth Monarchy representatives and Oliver Cromwell himself," Nisbet comments. "There was great joy in the hearts of most of the movement when they learned of Cromwell's interest, and he was even regarded as the Moses who would lead the way. Although there were some proto-populists who thought the government of saints should be chosen by ballot, there were many others who argued for a Sanhedrin of seventy impeccable earthly saints to be chosen by Cromwell, 'the great deliverer of his people (through God's grace) out of the house of Egypt.' "

The doctrine of predestination was linked to that of providence. Calvin, in the first edition of his *Institutes*, used the words "Predestination" and "Providence" interchangeably. But later, in the edition of 1559 (which is virtually identical to the 1561 edition Newton kept in his library at Trinity College), he separated the two terms, making providence the theme of the final chapters of Book One, dealing with God the Creator. He put off a discussion of predestination until Book Three. One reason was that predestination was restricted to humans and angels, whereas providence's realm was the whole of Creation. It applied to the universe in its entirety. In Cromwell's time, the austere doctrine of providence— the early Calvinist one that emphasized God's unalterable and eternal *diktat*—though unpopular in some university quarters was a force to reckon with, and it intruded into political life. High Calvinism was exploited as a means of welding marginal groups together at times of crisis and social upheaval. Once the worst of the crisis was over, the bleak, unfeeling nature of the doctrine was seen as unsuited to the gentler climate of the times. Though it was still proclaimed from many pulpits, there was a gradual recognition that it served no useful purpose. As early as the sixteenth century, Jacobus Arminius, a Dutch theologian, soothed an anxious laity by telling them Calvin was wrong: God's decree on the fate of individuals depends on whether they are truly repentant. It is not fixed for all eternity. So providence was at least partly in the hands of the human person. Arminius was a factor in the waning of undiluted Calvinism as the eighteenth century got underway. Once confined to an outsider, fringe culture, Arminianism became important in the shift from a "we happy few" theology of election and living sainthood for the lucky to a more inclusive and generous view of where the world might be going.

John Milton took to Arminianism in the 1650s, partly because he wanted a reasonable God, a God who made sense, in what was the birthing of the modern era. He did not feel comfortable with a deity who was all capricious power. He took the commonsense view that a person can love God only if there is completely free choice in the matter and we are not locked into a predetermined destiny.

Ultimately, the doctrine of predestination, with its seemingly irrational distribution of bliss and misery, faded into the worldly attitude of "life is unfair." Daniel Defoe is a striking example of a thinker of high intelligence unable to decide to what extent God intervenes, and how far humans are responsible for their fate. Defoe could take the high, religious road of warning that the plague, the weather, a financial disaster, are all signs of God's displeasure that the world is not proceeding according to his plan. On the other hand, there is no sense in taking such portents lying down. In the year 1703 a terrible storm, raging for a whole week, did appalling damage in England. Queen Anne, in a proclamation published in the *London Gazette*, accepted it as chastisement and spoke of the "crying sins of this nation." She suggested a fast. Bishops were asked to write special prayers. Defoe, however, trod warily in this matter. He himself had narrowly escaped being killed as the storm ripped a next-door house to pieces. He was tempted to escape from his own shuddering dwelling into the garden, but outside tiles were flying almost horizontally for a distance of some forty yards off his neighbor's roof. Defoe decided to "surrender to the disposal of Almighty Providence," and he resigned to die in the ruins of his house rather than meet certain destruction in the open yard.

But was the storm an act of correction by God for England's shortcomings? Defoe was ambivalent on this question. It suited his dissembling nature to perch on a fairly uncomfortable fence. He surmised that the seven-day tempest might be a reprimand for the shabby way the king had been treated, but described this cautiously as a "feeling" rather than a reality. Defoe was perhaps typical of his time. He could be severely moralistic in a conventional way, hot for "family values" and the influence of a strongly virtuous head of the home. In the first volume of his *Family Instructor*, for example, a little boy chides his father for making religion into something trivially social. Yet Robinson Crusoe, hero of Defoe's most famous novel, left his family to seek his fortune in the big world, and seems to feel no particular qualms about having done so. When down on his luck, in prison or being hounded by creditors—he was twice bankrupt—Defoe said he trusted in providence to restore him to better times. But it is clear he was kept sanguine by his own elastic and optimistic temperament.

Defoe was a transitional figure in an age of transition. The natural was edging out the supernatural. Domestication of the inscrutable was taking place, but the process was incomplete. No calamity prompted a more equivocal reaction—trust in God but keep your powder dry—than the plague, which struck crowded London when Defoe was a child of five. It lasted a year and a half and killed perhaps as many as 100,000 people. Defoe's *Journal of the Plague Year* (1722) was an account of the disaster. It reflects his own inconsistency by being part fact and part invention, a first-person report by a fictitious observer, "H.F." This narrator is also of two minds, flirting with superstition but then having no use for it. He makes plans to leave the infected city, but always finds some accident to prevent him: he cannot hire a horse, or else his servant has absconded. H.F. states that these frustrated plans must have been "intimations from heaven," since nothing happens without the permission of Divine Power. In that case, it was God's intention that he should stay. To do otherwise would be "flying from God." His brother, however, a widely traveled merchant, scoffs at such talk and tells him about the "Turks and Mahometans" in parts of Asia stricken by the plague. They were strong believers in providence and predestination who, thinking their fates had been decided unalterably before they were born, would walk without a qualm into infected places and blithely converse with contaminated people. As a result, they died off at the rate of ten or fifteen thousand a week, whereas the European or Christian merchants kept out of the way and largely escaped the contagion.

That throws H.F. into a frenzy of indecision. He spends an evening wrestling with his dilemma, torn between an intuition that his brother was correct and "the intimations which I thought I had from heaven" to stay put in London. Seeking a sign, he riffles through the pages of a Bible, stopping at the Ninety-first Psalm and the words: "Thou shalt not be afraid for the terror by night; nor for the arrow that flieth by day; nor for the pestilence that walketh in darkness; nor for the destruction that wasteth at noonday. A thousand shall fall at thy side, and ten thousand at thy right hand; but it shall not come nigh thee." At that moment H.F. decides to put himself in the hands of divine providence and stick it out in the plague-ridden city.

London, and H.F., do come through the horror—after a fashion. But it was not all the doing of providence. Human efforts play their part. Charity steps in, from the palace down, helping to keep the situation stable. The dead are taken away and buried. There is no runaway inflation and starvation is not widespread. The government does not collapse. Many people behave with decent restraint, though others resort to robbery and murder. And H.F.'s nightmare journey through the desolation of the plague prompts him to warn that just because a text in the Bible seems to offer protection against disease is no strong reason to take needless risks. "Upon the foot of all these observations," H.F. concludes, "I must say, that though Providence seemed to direct my conduct to be otherwise, yet it is my opinion, and I must leave it as a prescription, viz., that the best physic against the plague is to run away from it."

God, the Bible, prayer, divine intimations in the everyday incidents of life are constant themes in Defoe's writings. But in many cases they point not toward a sturdy faith but away from it, in the direction of worldly common sense and practical answers to life's predicaments. "The relative impotence of religion in Defoe's novels," says the literary historian Ian Watt, "suggests not insincerity, but the profound secularization of his outlook, a secularization which was a marked feature of his age—the word itself in its modern sense dates from the first decades of the eighteenth century." On the last page of Defoe's final book about Crusoe, *Serious Reflections During the Life and Surprising Adventures of Robinson Crusoe*, Crusoe laments the ebbing of religious devotion, which he predicts will not revive "till Heaven beats the drum itself, and the glorious legions from above come down on purpose to propagate the work, and reduce the whole world to the obedience of King Jesus." In a chilly aside, Crusoe adds that of this Second Coming, "I heard nothing in all my travels and illuminations, no, not one word."

Grand theological principles might, as the novel came into its own and the reading public expanded, be transmuted into literary devices. Divine providence, in works of fiction, might appear in the guise of "poetic justice." Defoe, in *The Further Adventures of Robinson Crusoe*, remarked on the curious fact that decent men often had the ill fortune to marry dreadful women, whereas some of the worst cads and reprobates— "scarce worth hanging"—boasted wives who were intelligent, careful with money, and hardworking. In real life, such glaring inequities are allowed to flourish; but in fiction, they can be remedied and often are. Wickedness does not win in the end, and virtue is more than simply its own reward. The critic Michael McKeon argues that the idea of poetic justice was in part a variation of the Reformation penchant for making a tight connection between the "ordinary," in the form of humdrum everyday life, and the theory of an active God furthering his plans. In that way the inscrutable mystery of divine providence could be made a little more accessible by virtue of its link to the "plot" of normal existence. But in the aftermath of the scientific revolution and increasing secularization, people seemed less willing to believe that the unfairnesses of life would be redressed at the end of the world. That sort of reluctance was stated by Robinson Crusoe in his farewell message. Better to make sense of the situation in the here and now, because the new generation was not so ready as previous ones to take on trust the existence of hell and its eternal combustion.

Samuel Richardson, in a postscript to his vast novel *Clarissa*, defines poetic justice as "another sort of dispensation than that with which God, by Revelation, teaches us." God places individuals on a sort of probation and "hath so intermingled good and evil as to necessitate them to look forward for a more equal distribution of both." Richardson notes that Joseph Addison, in an essay in the *Spectator* (vol. 20), called this a "ridiculous doctrine," since providence dispenses good and evil to all men alike. Addison also deplored the efforts of Nahum Tate, who rewrote *King Lear* to give it a happy ending, rewarding the good characters in this life rather than in the next. Tate's version of the play appeared in 1681 and was immensely popular until the nineteenth century. Tate deleted a quarter of the original text and had Lear live on happily with Kent and Gloucester after fighting off forces bent on murdering him. Cordelia survives unscathed and becomes an obedient wife to Edgar. "In my humble opinion," Addison wrote, this version of *Lear* "has lost half its beauty."

The term "poetic justice" was first used by the seventeenth-century critic Thomas Rhymer, who in an essay on *Othello* complained about the murder of the innocent and slandered Desdemona, remarking: "Is not this to envenome our spirits, to make us repine and grumble at Providence, and the government of the World? If this be our end, what boots it to be Vertuous?" Rhymer suggested amending the plot of the play, sparing Desdemona and having Othello "honestly cut his own Throat, by the good leave, and with the applause of all the Spectators."

Another "otherworldly" theme which took the downhill road to worldliness was that of the millennium, a central article of faith for Fifth Monarchy adherents. Millenarianism—the expectation, as previewed in the Book of Revelation, of the end of history and the inception of the thousand-year rule of Christ—was a belief entertained by intelligent, even exceptionally bright thinkers in the first half of the seventeenth century. Among them were Newton; John Napier, inventor of logarithms and others who studied biblical prophecies in a scientific spirit. Protestant study of the Bible was intended to rescue prophecy from mystery mongers and quacks and put it on a rational footing. Such researches encouraged the hope that the millennium was just a few years away. It would be preceded by the overthrow of the Antichrist. Puritan sermonizers took on a confident tone and a utopian tint. Great events were in the offing, perhaps arriving before the end of the century. The Civil War was seen as a prelude to the Second Coming, and people were urged to take arms on the parliamentary side to hasten its arrival. Newton's tutor, Dr. Joseph Mede, a professor of Greek at Christ's College, Cambridge, and a polymath of immense erudition, instigated a new theory of the last days, portraying them not as a time of terror and destruction but rather as the final stages of a process that would culminate in a Golden Age. God's plan for the redemption of humankind was thus made part of secular history, a history in which the powers of evil were thwarted by the cleansing force of the Protestant Reformation, which Mede described as "shaking off the yoake of the Beast."

Such a view began to take hold. For centuries it had been assumed that the Kingdom when it came would be not of this world, an exit from history. But during the seventeenth century there was a glimmer of optimism that human society here on earth could be improved to the point where it might plausibly be described as God's Kingdom. God was in fact already redeeming individual souls and the social order in readiness for such a terrestrial paradise. The Reformation, it was argued, by enabling the gospel in its pristine state to reach every person had effected a nearly incredible alteration in the human mind. An act of grace had supposedly produced a new kind of individual: more just, more truthful, and more charitable. Not perfect, but much improved. Such people were the raw material of a this-worldly millennium. New Protestant study of the scriptures highlighted the historical character of the Old Testament and the idea of a long-term scheme of redemption for humankind, with victory over Satan to be accomplished in history and here in our world. Joseph Mede was one of the leading scholars interpreting the Bible in this way. The eighteenth-century revivalist preacher Jonathan Edwards described God's program of redemption as a great "machine," its wheels rotating unstoppably and in perfect synchrony. Not all is harmony, however. The desired millennial state cannot be reached without armed struggle, military expeditions, and bloody conflict. In England, mainstream Anglicans and dissenting sectarians were in broad agreement as to the coming of an ultimate worldly kingdom of goodness. It was a patient kind of waiting, since many theorists of the millennium argued that it must arrive in stages, not all at once.

The historian Ernest Tuveson has made a case that the concept of the millennium, brought down to earth and into history, was one factor feeding into the notion that the United States has a divinely appointed mission as redeemer of the rest of the world. The term "manifest destiny" was not an expression of simple nationalism. It was, Tuveson narrates, the logical development of an argument, started earlier in England, that

God had a plan for universal salvation through history. In the America of the 1760s, after the French and Indian Wars, the idea began to circulate that the colonies were a "separate chosen people," given the task of completing the Reformation and setting up God's Kingdom on earth. John Adams wrote that the settlement of America was "the Opening of a grand scene and Design in Providence." In 1771, Timothy Dwight's anonymously published poem *America* spoke of the not yet independent colonies as glorious rulers of Empire:

> *Round thy broad fields more glorious* ROMES *arise,*
> *With pomp and splendour bright'ning all the skies;*
> EUROPE *and* ASIA *with surprise behold*
> *Thy temples starr'd with gems and roof'd with gold.*

The millennium had been translated into a this-world event. It was now to be made a New World project. In the last decade of the eighteenth century, David Austin, a biblical scholar and preacher, claimed to detect in the prophecies of scripture support for the idea that America was specifically appointed to complete the millennial task. Austin personified the Declaration of Independence as a "manchild hero," chosen to tread the beast of the Antichrist beneath his feet and bring in the promised Kingdom: "Follow him, in his strides, across the Atlantic! See him with his spear already in the heart of the beast! See tyranny, civil and ecclesiastical, bleeding at every pore!" Tuveson thinks we can see the American intervention in World War I as the accomplishment of this prophecy. He coins the phrase "millennialist nationalism" for such effusions, of which there were many. Colonel David Humphreys, a diplomat and protégé of General George Washington, held that America would in time "become the theatre for displaying the illustrious designs of Providence, in its dispensation to the human race."

The term "manifest destiny" was not invented until 1845. Tuveson believes that new possibilities of expansion for the young country "came into a kind of chemical combination with the general Protestant theology of the millennium, and with the already old idea of the destined greatness and messianic mission of 'Columbia.'" In 1858, an article in *Harper's New Monthly Magazine* entitled "Providence in American History" asserted that the basics of American democracy had developed from the idea of a Kingdom of God on earth. The "American Idea" was virtually a creation of the pure religion of Protestantism. The author spoke of a "national religion" that included millennial beliefs, with a capacity "to awaken the sense of Providence in the breast of a people." Herman Melville, in his novel *White-Jacket*, made a comparison between the United States and God's chosen people of Israel, stating that God had predestined "great things from our race," and he added that "The rest of the nations must soon be in our rear."

"Destiny," says Tuveson, became a secularized name for the religious vision of the earthly millennium, with America as the elected agent of that destiny. Even after the horror and waste of the Civil War, the myth of the Redeemer Nation retained a profound appeal, surfacing strongly in moments of crisis. When Woodrow Wilson made his tour of Europe in 1919 to win public support for the League of Nations, he said it was a moral obligation not to go back on the American soldiers who had fought and died in the war against Germany, so as to "make good their redemption of the world." Millennialist beliefs probably did not inspire the great decisions of American history simply by their own power. But Tuveson thinks they did affect national expectations about their outcome and results. And expectations based on the idea of a chosen people fulfilling God's plan are powerful stimulants to undertaking grandiose missions of redemption in history. That may help to explain why, in spite of setbacks and disasters, the United States plunges into enterprises such as the Vietnam War, the invasion of Afghanistan, the forced "liberation" of Iraq. "The United States' 'destiny' to save the world, is the product, not of political idealism and observation, but of prophetic beliefs."

CHAPTER FOURTEEN

The Private and the Public

If God wanted his tracks discovered, wouldn't He have
made them plainer? Why tuck them into odd bits of
astronomy and nuclear physics? Why be so *coy*, if you're
the Deity?

—JOHN UPDIKE

Not long after the glorious revolution solidified the new
stability of society and a cooler climate of religion emerged, churchmen
noted with alarm that something more exciting than theology had
gripped the English, especially the London, psyche: money, and the pleasures, luxuries, and entertainment money could buy.

"Is it to be wondered," asked Edmund Gibson, bishop of London,
from the pulpit, "that the people should be indisposed to attend to anything serious, or that they grow sick of religion, which has no comforts
for them; that they fly from the church and crowd to the playhouse?
They are tired of themselves, and their own thoughts, and want to lose
themselves in company from morning to night."

London in the seventeenth century was becoming the largest industrial center in Europe. England was entering on an extended period of
remarkable growth and prosperity, buying and selling on a global scale,
creating a demand for the exotic and opulent as well as for humdrum
daily necessities. The reach of England's commerce stretched from

Morocco, Russia, and Persia to Venice and the East Indies. A more affluent consumer was demanding a more luxurious range of products. A rising merchant class was moving into positions of influence and power in London's government. There emerged a "merchant political elite" based largely in the Levant Company, which did business not in the parochial enclave of northern Europe but in southern Europe, the Mediterranean, and the Near and Far East.

St. Paul's Cathedral might still be the center of London's worship. But an institution of Mammon was emerging as a rival attraction: the Royal Exchange, hub of commerce where merchants congregated from all over the world. The godly sophisticate Joseph Addison wrote ecstatically in 1711: "There is no place in the Town which I so much love to frequent as the Royal Exchange. It gives me a secret satisfaction, and, in some measure, gratifies my Vanity, as I am an Englishman, to see so rich an Assembly of Country men and Foreigners consulting together upon the private Business of Mankind, and making this Metropolis a kind of *Emporium* for the whole Earth." Addison especially admired the foreignness of the goods on display. "The Food often grows in one Country, and the Sauce in another. The fruits of *Portugal* are corrected by the Products of *Barbadoes*: the Infusion of a *China* plant sweetened with the Pith of an *Indian* cane. The single Dress of a Woman of Quality is often the Product of an hundred Climates."

England's commercial and industrial classes, individualistic in their outlook, now exerted more economic and political influence. Most important, society had reached the tipping point between a largely religious and pious society and a new order based overtly on money, power, and individualism. In London, amounts of money huge by the standards of the time were won or lost in stock speculation in the boisterous atmosphere of Exchange Alley or in coffeehouses. Often the commodities traded in these transactions had the flimsiest basis in reality. As the saying went, stockjobbers hoped to "sell the bear's skin before they have caught the bear." Gambling fever had crossed the Channel from France. Soon Covent Garden, where Londoners entertained themselves after dark, had thirty gaming houses. There were public lotteries with offices all over London. The government had control of taxation and over how the revenue was spent, making possible the commercial and financial revolutions and the explosive growth of commerce. A permanent national debt was created. Parliament regulated trade and chartered commercial enterprises, easing the flow of capital. It cracked down on monopolies, most notably the Hudson's Bay and Royal Africa Companies, responding to the wishes of influential London merchants with links to Whig politicians. The government's soaring debt was propped up by borrowing in the Exchange, a practice that led to the scandal of the South Sea Bubble.

England became home not to the dawning of a millennialist paradise but rather to pockets of wild affluence amid a landscape of disgraceful hardship. It was clear that the optimism of the Newtonian churchmen had been overdone. The timetable of apocalypse needed adjusting or else should be abandoned altogether. "Take heed of computing" the date of the millennium, warned John Owen, a member of Parliament. "How woefully and wretchedly have we been mistaken in this." Those who clung to the apocalyptic vision of the millennium tended to be aggrieved at the indifference of the official Church and the state of society in general. They "despaired of the meaning of history." Only the end of the world would give that meaning back. Then and only then would there arise the perfect society, perfectly just, perfectly moral. This millennial fantasy was in many ways an indictment of the establishment for its failure to create such an ideal world. It was an embrace of a supernatural utopia as a last resort, given the failure to bring in a mundane one.

Simultaneously with the new focus on the human, the term "the public good" began to appear increasingly in sermons, pamphlets, and various writings. Richard F. Jones, in his classic study of the rise of the scientific movement in England, links the new science and the altered attitudes toward knowledge it promoted to the notion that the advance of physics and chemistry is inextricably linked to the progress of the general human welfare. Sir Francis Bacon had taken a strongly utilitarian view of science and its potential for enabling humans to command the forces of nature. His sense that science could grant humans new powers over their environment was not shared by all Newtonians, for sure, but it was an article of faith for Baconians. Jones notes the tendency for "pure" science, one richer in theory than in practical applications, to be associated with the "proud speculative man in contrast to the humble experimenter working for the good of mankind." New scientific interest was shown in agriculture and the manual trades. The

Baconians held a certain hostility toward the ancients because of their reliance on abstract models and mathematical artifice rather than on experiment and observation. Bacon came to be seen as the great antagonist of the ancients and champion of the moderns, and his stress on the public good had something to do with this. Esteem for mere thinkers declined at the same time that it rose for artisans and people who worked with their hands. Farmers and mechanics enjoyed a new respect for being in touch with the "real" world, and for operating in the realm of the useful. The Royal Society welcomed such types into its membership. Landed aristocrats and gentry would extend the hospitality of their dining tables to anyone with news of up-to-date agricultural methods and machinery.

Puritan Reformers, who had now taken on the duty of promoting Bacon's gospel of "The Useful," found sanction in another value which, if not new, became more potent than ever before, namely, "the public good," or the welfare of society. Here, says Jones, was laid the basis for a motive so ardent it dominated educated opinion. Naked materialism needed to be decently covered up, and humanitarian ideals were as good a costume as any. The welfare of the poor became an important justification of a materialistic program.

In his *History of the Royal Society,* Thomas Sprat repeatedly asserts that science, unlike unmediated contact with the divine, unlike mysticism, is a *public*, collaborative enterprise. Finding new technologies requires the efforts of many people sharing what they have discovered, not a single mind in secret conference with the Creator. Newton's intense solitariness, his suspicious nature, his jealousy and fanatic need to be first, were an aberration at the birth of modern science.

Indeed, the Society's very purpose was for various investigators to discuss or to conduct experiments as part of a like-minded company. At first there were weekly meetings of "inquisitive" people in Oxford, who paid dues to finance experiments. They congregated at the house where a man was available to grind glasses for microscopes. All talk of religion and politics was banned, conversation being restricted to physics, anatomy, geometry, astronomy, navigation, magnetics, chemistry, mechanics, and other scientific topics. It was dubbed "the Invisible College," expressing the cooperative spirit of the meetings. According to Sprat, the principal aim of this collection of inquiring minds was "that they might enjoy the benefits of a *mix'd Assembly*, which are largeness of observation and diversity of judgments, without the mischiefs that usually accompany it, such as confusion, unsteadiness and the little animosities of divided Parties." An egalitarian spirit prevailed, because it was the artifice of experiment, not the acuteness of someone's commentary on it, that was held in respect at such meetings. In vain would any member hope to make himself conspicuous by the subtlety of his wit. Disagreements could not be disruptive, since nature may have different ways of bringing about the same result, all equally good, whereas "they that contend for truth by talking, do commonly suppose that there is but one way of finding it out."

There was a free and easy atmosphere, suited to a collection of open-minded people. Their purpose, Sprat said, was to "heap up a mixt Mass of Experiments, without digesting them into any perfect model." Their discussions followed a "roving and unsettled" course. Sometimes they let themselves be guided "according to what any foreigner, or English Artificer, being present, has suggested: sometimes, according to any extraordinary accident in the Nation." The subjects of their researches were also multifarious: "experiments of rarefaction, refraction, and condensation: concerning the cause, and manner of the Petrification of Wood: of the Loadstone: of the Parts of Anatomy, that are yet imperfect: of Injections into the Blood of Animals; and Transfusing the blood of one Animal into another: of Currents: of the ebbing, and flowing of the Sea: of the kinds, and manner of the feeding of Oysters: of the Wonders, and Curiosities observable in deep mines."

It was about the time that the *History* was being written that today's historians detect a new significance surrounding the term "public." The word had been used in contrast to "private" at least since Roman times, of course, making a distinction between the state and its properties and the citizens in their homes. But "the public," in the sense of a collection of private individuals whose opinion counted, was an early modern development and was not in general use until the eighteenth century. The novel and the theater, as increasingly commercial ventures, each had its public, which decided whether or not a particular publication or production had merit. Especially in England, where censorship had virtually collapsed after the revolution, where the party system led to uproarious debates in the House of Commons, and where there was an explosion in political pamphleteering, public life was entering a remarkably strenuous phase. Journalists were growing more openly impertinent, playwrights more irreverent, and the coffeehouses served as a stage for wits and atheists to publicly deflate the pretensions of public eminences. In Cromwell's New Model Army, England's first standing army, soldiers away from home for the first time aired their grievances and comments on national affairs in an uninhibited, open arena. It has been said that English public opinion was born around the campfires of the New Model Army.

New venues opened for the expression of collective interests. Taverns, once shunned by the middle class as resorts for plebians and social outcasts, became meeting places for political organizations and debating societies. They had a role to play in the development of party politics. At election time, supporters of a candidate were given tickets entitling them to free liquor and food, or even a bed for the night at an inn. Later, Charles Dickens satirized this practice in his account of the Eatonswill election in *The Pickwick Papers.* Coffeehouses had a huge vogue by the beginning of the eighteenth century, thanks in part to the political turmoil of the time and the boom in political journalism, all fodder for uninhibited argument. By the coronation of Queen Anne there were about five hundred coffeehouses in London. In the 1660s, radical ideas about setting up a republic in England and heretical tilting at orthodox religion went on at the famous Turk's Head, frequented by Samuel Pepys. At Oxford, where the first English coffeehouse was started, the "Vertuous and Learned" precursors of the Royal Society would gather at a house near All Souls to quaff coffee at a penny a dish. After the Society was formed, some of its fellows congregated at the Grecian coffeehouse in London's Threadneedle Street.

The government sometimes used the loose public ambiance of the coffeehouses for its own propaganda purposes. After the Restoration, they had become an informal substitute for suspect political clubs that had been shut down by the authorities. Impudent news sheets were distributed to the customers, tailored to their own particular leanings. The government shilly-shallied over banning the seditious ones; many in power worried about their potential for spreading war rumors. In the end, however, it decided to let them flourish, occasionally exploiting their immense potential for publicity. In 1665, Samuel Pepys was asked to go into the coffeehouses to put about a story that the Dutch were abusing English seamen, where the tale could "spread like leprosy."

Elections, too, were more numerous. A new law made it compulsory that elections be held every three years at the latest. The historian James Melton has described how this created a public accustomed to taking an active part in politics. The Civil War had played a significant role in stimulating a larger and more vocal public, better informed about what the politicians were up to. In many ways, the Glorious Revolution had helped to "deprivatise" society, especially in London. The War of the Spanish Succession, a wildly expensive venture, meant the crown was hardpressed for money and had to summon Parliament annually, ensuring fractious debates and leaks of information to the press.

Such openness can be exaggerated. When the Whigs took power in the first half of the eighteenth century and the Tory Party went into eclipse, the intensity of party warfare faded and elections were held less often. The crown exercised its top-down powers by means of patronage, offering tempting benefits in the form of jobs, contracts, church livings, and commissions in the army to its supporters, while a ruling class of aristocrats helped to buttress royal authority. Along with a decrease in the frequency of parliamentary elections came a drastic expansion in the number of uncontested seats. But in spite of this cooling off of two-party rivalry and the rise of a parliamentary elite, the "rituals" of election politics drew large numbers of nonelite people into their compass.

Candidates used the press to publicize meetings and dispense propaganda. Thousands of handbills were distributed and dinners were hosted for supporters by the hundreds. Significantly, it was no longer safe for a candidate for a contested seat to be aloof and private, as if reluctant to disclose himself to the rowdy and uninhibited public. James Melton tells the story of William Hay, an election agent for the Duke of Newcastle in the 1730s, who warned the duke that one of his parliamentary clients might lose the coming election due to his reluctance to get to know the voters in person: "He has not been around the Town since he went with your Grace, nor I believe asked a single man for his vote; and I am firmly persuaded that half the voters that have been lost have been lost by his unpardonable negligence; the people are affronted by it."

Another factor in the growth of a public with opinions of its own was a revolution, alongside the scientific revolution, in the development of literacy. There was a big bang in the production of reading materials, including books and newspapers, periodicals and almanacs, not all secular by any means. The number of Bibles and sermons increased, though as a declining percentage of the book market as a whole, since this was a revolution in preferences. At the height of the Newtonian period, religious works amounted to about half of all books published. At the close of the eighteenth century, they made up less than a quarter. As for readerly habits, there was a shift from "intensive" reading of a small number of texts, studied closely and repeatedly, to "extensive" reading of a much wider spectrum of materials, a sort of popular echo of the impressive eclecticism of the Royal Society's interests.

Intensive reading, characteristic of households in the seventeenth century, was usually not a source of information about public matters, but a private activity within the family in which books were used to reinforce and restate one's religious faith. That practice did not die out in the next century, but among educated people in the towns there also grew up a more portentous taste for other kinds of literature, including the novel, which were read less reverently. Skimming came into fashion. Dr. Johnson, whose shelves groaned with the weight of about three thousand books, thought—or pretended to think—that to read a book from cover to cover was a most immature and pedestrian thing to do. Sometimes he simply read the cover. Later, in the Victorian era, Charles Darwin became notorious for taking a book into his hands and ripping it into two parts, destroying the binding, to make for easier handling.

In this way print became "demystified." In rural areas, books were regarded almost as icons, as sacred objects, even possessing occult powers. One swore on the Bible to tell the truth, or placed it near a sickbed to work its healing influence. But among the urban literates, a book was just a book, and one need not be too overawed in its presence. The "desacrilisation and commodification of books," James Melton notes, "in turn fostered a less deferential attitude toward print. Like consumers choosing from a variety of wares, extensive readers were more apt to make critical judgments and comparisons." That inevitably created a literature that was not only demystified but deprivatized as well.

Since the eighteenth century, the public dimension of society has never ceased to expand at the expense of the private. This new balance between public and private presented something of a quandary for confessional religion: how far should the Church go in moving with this powerful social and intellectual trend? Since Newton, there has been a great temptation for religion to go into partnership with science, perhaps the most public and officially protected sector of modern life. But the liaison is apt to look markedly one-sided. Scientists can on occasion be remarkably offhand and careless in their public statements relating their discoveries to theology's chief interests. This suggests that such analogies, while interesting and thought-provoking, are not of fundamental, pressing value to science. It seems that theology needs science more than science, as a search for "truth," needs theology. Robert Russell, a physicist and founder of the Center for Theology and the Natural Sciences at Berkeley, thinks religion is incapable of making its moral claims or providing spiritual comfort unless its intellectual arguments are credible. Modern theology sometimes gives the impression it is snatching at whatever modest opportunities scientists can put its way. The situation is not quite as bleak as when Bertrand Russell, with his usual cruel succinctness, wrote that "Theologians have grown grateful for small mercies, and they do not much care what sort of God the man of science gives them so long as he gives them one at all." Linking up with exciting and highly publicized successes in astronomy and cosmology gives religion a public dimension and a "seat at the table" with an enterprise that today is largely government-sponsored. It is worth noting, however, that Methodism, which accentuates the personal element and "warmer" side of God, never became tremendously excited about new theories of the universe. Religious thinkers have regained their footing now. But it is still true, as it was in Newton's day, that scientists are inclined to give us God the Scientist, or more specifically God the Physicist, not a being who listens to petitionary prayer and has a plan for the world. Not to mention the doctrine of the Trinity.

Theologians were particularly intrigued by the Big Bang theory of the origin of the universe, first proposed in the 1930s. The theory was sufficiently imprecise to lend itself to speculation as to whether it supported the biblical account of a moment of creation by an Author who made nothing into something. The physicist James Jeans, in a book written for the general reader, suggested that matter was created by high-energy photons crystallizing into electrons. He added, in a throw-away line, "we may think of the finger of God agitating the ether." Jeans himself called this metaphor "crude." He eventually backtracked to the point where he speculated that the universe may be only a "thought in the mind of the Creator," in which case it would be pointless to even talk about the Creation.

One of the earliest proponents of the Big Bang theory was a Belgian Catholic priest and physicist, Georges-Henri Lemaître. His ideas were partly based on the uncertainty principle of quantum mechanics, which avoided the dilemma of an infinite regress. In his account, the universe evolved from a single primeval atom. Lemaître wavered, Hamlet-like, in deciding whether to discuss the role of God in a letter to the journal *Nature* concerning his theory. In the typescript of the letter he impulsively penned a paragraph in which he celebrated the fact that the Big Bang effectively protected God's privacy and left him discreetly unobserved: "I think that everyone who believes in a supreme being supporting every being and every acting, believes also that God is essentially hidden and may be glad to see how present physics provides a veil hiding the creation." But before sending the letter off to press, he crossed out this seemingly innocuous comment. In fact, Lemaître believed religion is first and foremost about saving one's soul; it is not a window onto nature. He stated that the doctrine of the Trinity is "much more abstruse" than anything in the theories of relativity and quantum mechanics. Lemaître thought the Big Bang model left the unbeliever free not to believe in any transcendental being, and at the same time prohibited anyone, believer or agnostic, from getting too familiar with God. It was a prophylactic, a barrier to prevent scientists from discovering too much about the workings of the divine.

Loose talk by scientists has sometimes warred against this commendably guarded view. Men and women of science can get familiar with God without a qualm. During the 1980s, a team of astronomers at the Goddard Space Flight Center in Maryland, set out to investigate why galaxies form in what should be a smooth, unlumpy mix of particles. They put a satellite into space named COBE, or Cosmic Background Explorer, to measure the background radiation left over from the Big Bang more accurately than ever before. COBE was to look for tiny ripples or imperfections that might act as gravity attractors, enticing particles to coalesce and form a large celestial body. To the delight of NASA, which sponsored the project, COBE found not only wonderful smoothness in the radiation, supporting the Big Bang theory, but also slight temperature fluctuations, "hot spots," millions of light-years across.

At a triumphant meeting of the American Physical Society in Washington, D.C., in April 1992, one of the scientists on the COBE team, George Smoot, let euphoria get the better of prudence. Smoot had invested years of work on this project. In the 1970s he had tried to look for cosmic lumpiness with balloons. These efforts regularly ended in disaster. The balloons that did land intact failed to show any irregularities. In a rash moment Smoot remarked of the COBE result: "If you're religious, it's like seeing the face of God." His comment annoyed scientists as well as theologians. Ironically, the discovery of an *imperfection* in the structure of the cosmos was being recruited as evidence of a God who in traditional theology was held to be perfect. Balky scientists pointed out that the COBE result would have been more exciting if it had *not* found temperature fluctuations. Instead of merely confirming a popular theory, it would have set cosmology back on its heels.

Lemaître was a trained theologian and a scientist who was more comfortable with the actualities of nature than with the flighty constructions of mathematics. He had qualms about going public with an idea associating science with his faith. Others were not so reticent. Edward Milne, a leading astrophysicist in the years before World War II, introduced a theory comprising two different time scales, in which the laws of nature alter with time and the universe has a genuine history and therefore a clear meaning. Milne unblushingly linked his theories to the doctrine of God's omnipotence, as if centuries in the checkered career of that principle were not a warning to be cautious. At a meeting in Brussels, he announced that the evolving universe must be infinite. He talked about "evolutionary experiments," rather in the manner of a science fiction writer, adding: "If evolution consists in the occurrence of mutations, there would be something little, and as it were pettifogging in the application of the evolutionary process to a finite universe, in which only a finite number of evolutionary experiments could be practiced. It would be to put Deity in a straitjacket. On the other hand, in creating an infinite universe, we can say that God has provided himself with the means of exhibiting and practicing his own omnipotence." There are uncomfortable echoes of Pope Urban VIII in that sentence.

Big risks are attached to moving religion, with its unique components of privacy and individual experience, into the public and secular arena. Theologians and scientists both are apt to make fools of themselves. Prayer, supposedly a core element of faith, is essentially a private affair. Today, many believers do not expect to have mundane requests granted by petition in this way, as one would call on a politician or an employer to grant a favor. Rather, it is a means of opening mind and spirit to a divine presence. But the unworldly and personal, as we have seen, have a habit of metamorphosing into the worldly and public. It happened to the doctrines of divine providence, predestination, and millennial utopias. These were reinvented and became the Idea of Progress, the idea that "life is unfair," and the improvement of the human condition by science. Moderate churchmen during the turmoil of the English revolution welcomed science as a partner in the cause of restoring public stability and fending off the radical, subversive sects. Calming down the dissidents, it was said, would put society into a suitably ordered state to receive the Second Coming. The idea of a universal science, secularizing Luther's dream of an expanded knowledge of religion, was a millennial concept.

A mistake made by Edward Milne and others was to carelessly fail to make a clear distinction between the physical *origin* of the universe, which is purely a matter of science, and its creation by a supernatural being, who presumably has made it for a purpose that includes the destiny of its human inhabitants. The Vatican was only too keen to gloss over this difference at the time. Pope Pius XII declared in his 1951 address to the Pontifical Academy of Sciences that modern science "has confirmed the contingency of the universe and also the well-founded deduction as to the epoch when the world came forth from the hands of the creator. Hence, creation took place. We say, therefore, there is a Creator. Therefore God exists!" It was a stunning surrender of faith to physics.

The pope's statement offended a number of nonreligious scientists, some of them supporters of the rival steady-state model of the universe, which left no room for a moment of creation. Physicists were not the only ones bothered by Pius's statements. Theologians suspected he was repeating the old error of tying the truths of revelation to science's apron strings, as had occurred when the Church unfortunately adopted

Newton's theory of a watch-winding God. Those strings, they knew, are prone to fray and unravel. Scientists who backed the steady-state theory were incensed to see the majesty of the Church put at the disposal of something they believed to be mistaken. Lemaître urged the Vatican not to muddle truths of revelation and scripture with scientific theories, which must always be treated as provisional. The pope complied. His next address to astronomers in Rome only a few months later avoided mention of religion in connection with the Big Bang model. It was left to Pope John Paul II in 1988 to make a bold declaration that Christianity is its own justification and "does not expect science to constitute its primary apologetic," thereby freeing it of any and all responsibility for worldly scientific advances.

What Lemaître was doing, of course, was telling Pius that science should be prized not for what it reveals about the Creator but for its affirmation that God cannot be known through the lens of physics and astronomy. Such human tools are in fact just new forms of idolatry, images of the divine all too easy to mistake for the divine itself. The new quantum theory might be abstruse and foreign to common sense, but it is not as strange as God. Be glad that is the case, was Lemaître's message. And be content with it.

There are those today who, in a very different way, are urging religion to take on a higher public profile, to move even farther away from its private character. Robert Drinan, Jesuit, former member of Congress, and now a professor of law at Georgetown University in Washington, D.C., has argued for setting up a world tribunal to monitor and protect freedom of religion. It is odd, he thinks, that the United Nations, with its elaborate machinery for enforcing human rights, has no mechanism whereby people who feel victimized on account of their religion can seek redress as a basic entitlement. The reason, in his opinion, is that religion today is thought to be too private a thing to be under the control of a world body. To some extent, religion is off the radar screen of a society that has tacitly decided that the subject is not a public concern and has nothing to do with supposedly secular politics.

"In a sense, the Catholic Church at the Second Vatican Council recognized the forces of secularism and gave up its seat at the table of governmental affairs," Drinan maintains. "It settled for assurances of tolerance and nondiscrimination from government forces, and this is now the stance of every religious group in Western culture. This will safeguard against some cases of open persecution, because churches are no longer the rivals of the government. A truce has been reached, with religion disarmed—and, some would say, made to some extent irrelevant. Many devout Christians are so content with the privatization of their faith, so approving of the new status and role of religious organizations, they actually consider the Church made a monumental mistake when it accepted the patronage of the Emperor Constantine in the year 315 CE."

Behind the ambivalence or quiet hostility to religion in the minds of millions of people, in Drinan's view, lies the desire, perhaps the hope, that international law will increasingly make religion a wholly private matter and stop it from taking an active role in world affairs. Many experts in international law are of the same mind, arguing that the advancement of human rights would be better off if it were free of religious influences. But this is not a healthy state of affairs. There is a certain kind of virtue that commonly radiates from devout believers and can only enrich governments, in which virtue is not always a conspicuous factor.

It may be that private religion is unable to persist indefinitely without a public component. That could explain why in our own time, in the countries of the old Soviet Union, there was no massive crackdown on personal devotion, quietly lingering on the margins, as long as it stayed on the fringes as a harmless eccentricity, playing no real part in the Socialist system. The English theologian Donald Cupitt sees something of that attitude in the Christian West today. Marxism made a very clear distinction between the old, medieval concept of a religion so overarchingly public it could not be confined to state and pulpit but took in the whole, ornate, and richly structured cosmos, and a private religiosity that cuts the umbilical cord attaching it to the world. A surreptitious version of that devotion could even flourish in Stalin's gulags. But it is wrong to say religion was "tolerated" by the Soviet regime. It was permitted to eke out a meager existence only because public religion had been completely cauterized from what was an intensely public regime: the remainder was regarded as being of no consequence.

Something similar has happened in other countries of the West, in Cupitt's opinion. Some vestiges of the old public reality of religion remain there, but one cannot deny that it no longer exists as an effective force. As religion gets squeezed out of the public domain, it also becomes less and less operative in the private realm and not as effective in sustaining personal morality. A good deal of private religiosity still persists, often of a whimsical and exclusive kind, but it makes little impact on the rest of us. The reason why it is insignificant, Cupitt thinks, is that private religion needs, and must presuppose, a public dimension if it is to make a difference in the world.

"In a curious way," he says, "Christianity seems to *need* a secular State to contrast with the Church, secular life and natural virtues to contrast with the religious life and the theological virtues, secular love and knowledge to contrast with its own sacred love and sacred knowledge. As a woman needs a husband to cherish, to improve and to rail against, so the Church needs a secular tradition. The secular spouse with which Christianity has been in this curious love-hate relationship has sometimes been classical philosophy, sometimes the Emperor, sometimes the man of letters, and, more recently the men of science. Christianity just happens to be like that."

CHAPTER FIFTEEN

Cutting God Down to Size

There cannot be a personal God without
a pessimistic religion.
　　　　　—CYRIL CONNOLLY

THERE WAS A TIME, as the new science was getting up a full head
of steam, when radical politicians and others were, in the words of
Christopher Hill, out to "cut God down to size." That outlook existed
side by side with a curious move within Protestantism best described as
militant unreason, a subversive project which took the form of using the
Bible to discredit prevailing norms. It borrowed from Luther's teaching
that God is beyond the reach of human understanding, which had been
tarnished and debased since the Fall of Adam. Given the mortal mind's
stunted gropings after an unobtainable truth, the Bible, divine revelation, was its only source of saving knowledge.

The purveyors of militant unreason regarded anything without a
proper basis in scripture as suspect. Popes were not mentioned in the
Bible. Nor was the Trinity, except in words that could be taken in more
than one sense. The absence of any mention of the doctrine of the
divine right of kings was used by partisans of the Glorious Revolution
to justify switching sovereigns. For some, one of the piquant attractions
of the "plurality of worlds" thesis, the idea that the universe teemed with
stars and planets containing intelligent life, was that not a word is said

about it in the Book of Genesis. Dissenters used this device to discredit the system of tithing for parish priests. But once exploited as a means of radical attacks on Church traditions, the Bible could then be used, not exactly for what it omitted to mention, but for its openness to alternative readings and slanted interpretations aimed at rocking the political boat. Scripture could be quoted to support a case, or its absence in the pages of scripture given as a reason to condemn it.

A person could construe sacred writings as concordant with the needs of society or with the cries of agitators anxious to knock society back on its heels, according to choice. Milton, apt to quote the Bible to unsettle doctrines not to his taste, endorsed such a practice, saying that "no ordinance, human or from heaven, can bind against the good of man." That includes not only his spiritual but also his physical and worldly good. If the Bible seems on the surface to go against that precept, then it must be understood in a different sense. The parliamentarian Henry Parker insisted that if a secular law is injurious to social justice, then "we ought to be very tender" how we try to reconcile it to God's law, or acknowledge God as the author of any edict "which man reaps inconvenience from."

In Christopher Hill's opinion, this sort of thinking not only tended to remake God in man's image (Voltaire had said that if God made us in his image, we have certainly returned the compliment) but implied that God must be taken down a peg, reduced in scale. Science in its great early blossoming could be a prop to faith, perhaps more of a prop than was good for either science or faith in the long run, but it was another matter with the rising school of biblical criticism. This led to skepticism and atheism more readily than science did. Selective interpretation of scripture was a common practice with dissenting sects, forcing the Bible to respond, willy-nilly, to pressing social needs. If the authority of the Bible were to give place to that of the individual conscience, then it was only a short step to saying we must decide and act according to our *social* conscience. In that case, God has devolved into a social concept.

What particularly aroused Luther's scorn was the idea that one could, by pressing reason to the farthest extreme, take this exotic, otherworldly, or divine being and domesticate him, make him more like us. He insisted that God does not behave like anyone we know. Theologians who tried to housebreak the deity were treated harshly. Luther called reason the Devil's Bride, the Whore, the Fool. And he did not hide the fact that when truth is *not* domesticated, it can be difficult to live with. The Reformation took two things with deadly seriousness. One was God's unutterable majesty and dominion, and the other was sin. The first implies the second. If the Author is so immensely impos- ing and dreadful, then sinning against him is a matter of huge conse- quence, something that cannot be lightly disposed of by confession, indulgences, or good works.

The debate over how active God might be in dealing with his cre- ation ran well into the Age of Reason and beyond. It revolved crucially around the question of to what extent God is "made in our image." Is he a mere Maker and Designer, skilled in geometry and remote from the hurly-burly of ordinary life? Or is he actively involved, doing what is good for humans, because he is like humans? This dilemma has never entirely disappeared from the Western consciousness, and it has been a spur to periodic attempts to shrink God to human scale, to de-divinize him.

In the Old Testament, especially in the Psalms, the splendors and terrors of nature are noted, but there is little sense of God playing an active role in its courses. Instead, the Psalmist seems more impressed with God's supremacy and the idea of divine order. Psalm 104 (2–5) celebrates a deity

> *Who coverest thyself with light as with a garment; who stretchest*
> *out the heavens like a curtain.*
> *Who layeth the beams of his chambers in the waters; who maketh*
> *the clouds his chariot: who walketh upon the wings of the wind.*
> *Who maketh his angels spirits; his ministers a flaming fire.*
> *Who laid the foundations of the earth, that it should not be*
> *removed for ever.*

Psalm 29 (1–5) urges:

> *Give unto the Lord, O ye mighty, give unto the Lord glory and*
> *strength.*
> *Give unto the Lord the glory due unto his name; worship the Lord*
> *in the beauty of holiness.*

> *The voice of the Lord is upon the waters: the God of glory thundereth:*
> *the Lord is upon many waters.*
> *The voice of the Lord is powerful; the voice of the Lord is full*
> *of majesty.*
> *The voice of the Lord breaketh the cedars; yea, the Lord breaketh*
> *the cedars of Lebanon.*

God is very real in these verses: he can exert his power over nature in the most violent fashion, and he is certainly to be praised. But he is also something so tremendous as to make human beings seem insignificant, and he does not act so much as dominate and assert his magnificence. There is a hint that, in certain circumstances, he might be an enemy. But of course God does act in the Old Testament. He supervises the fortunes of his chosen people, the Israelites, and intervenes directly when that is needed. It is almost as if God had two identities, and it depends on the temper of the times which one is emphasized and which minimized in any given circumstance.

In our own day, theologians have not come down unanimously on the side of the active God. James Gustafson thinks a "systematic" discrepancy between the God of action and the God of inaction has existed in the past and still exists now. It harks back to classical times. Cicero wrote that whereas some people assume the gods have no concern whatever for the affairs of mortals, others say they are intent on providing for human welfare. He summed up: "There is no subject on which there is so much difference of opinion among the learned and the ignorant."

Luther and Calvin held the two kinds of divinity, caring parent and impersonal power, in balance, because they could then move from the Old Testament breaker of cedars to the gentler New Testament atmosphere. Gustafson puts the case for retaining some of the Old Testament austerity, which surfaced briefly in the decades after Newton. He suggests that a more impersonal profile of God may be preferable to the warm and cuddly one, which tends to foster forms of piety that are radically individualistic, human-centered, and utilitarian. Better an impersonal deity who is the source of human good, but who does not guarantee it.

If God is seen as a benevolent rich uncle who hands out favors to prayerful supplicants, some of them quite trivial, the civilized values of order and constraint that give life a needed structure, with rules and norms, are marginalized. One of the major shifts in Protestant thought in our time, Gustafson thinks, took place as a result of the exclusive, or at least primary, focus of attention on the idea of the God Who Acts. It broke free of the spell of science to construct a God whose chief importance is not that he made the cosmos, his purposes being revealed in the order of creation, but rather that he operates in history. This suggests that his entire work is taken up with securing human welfare and redemption, a notion Gustafson thinks deeply misguided. Liberation theology, that curious bonding between Jesuits and Marxists to support left-wing insurrections, is a prime example of this change of emphasis. Science does supply some boundary conditions and constraints, which is why we should take science seriously. Absent those checks, the "rules" of life are apt to be interpreted to suit the needs of the moment. Belief in an ordering, impersonal God gives more authority to laws than does a radically personal one, and to institutions that provide stability for a society based on principles of justice.

It all depends on how you define God, and to a lesser extent, how you regard the nature of the universe. It was a tenet of liberation theology that humanity is "evolving a more correct idea of God." Some of today's scientist-theologians would have us believe the universe is a "home for humans," fine-tuned in its basic structure to permit the emergence of man. Such a sentiment implies a God whose whole intent is the welfare and flourishing of the human species. Gustafson finds this to be vanity verging on narcissism. Our world is not a celestial Garden of Eden designed to accommodate us in perpetuity, he argues. It had a beginning, it took millions of years to evolve, and one day it will all come to an end. "If it were not for that knowledge," he writes, "one might live contentedly with the traditional Western assurance that everything has taken place for the sake of man." Yet theologians such as Jürgen Moltmann blithely interpret the Christian story of resurrection as proof that nature, rather than having a cold dark terminus when the Sun burns out, is open to a "new creation."

Is it God's role to cater to our needs? Perhaps God's power and order limit human activity and "demand" recognition of principles and boundaries. Remember Bonhoeffer's prison musings, that the modern world treats such boundary conditions as sin and death as if they do not exist.

Science can be a reminder that they do exist. It was the "naïve anthropocentrism" of a religious culture that prevented brilliant scientists from seeing the clear significance of their own discoveries of the geological layers of the Earth, and of the extinction of species, blinded as they were by the belief that humans were the point and purpose of a divine plan.

Seeing ourselves in that light tends to conjure up a God so singlemindedly bent on looking after us that he takes on a human, a very personal disposition. And that may lead to treating the nonhuman parts of the universe with cavalier heedlessness. If God is ruler of the entire creation, then humans are no longer the sole measure of value in that creation. And they are not the only reason the universe came into being. Gustafson goes further than this. He proposes that what we decide is good for human individuals, or for humans as a species, may not agree with the ordering purposes of God, as far as those purposes can be discovered. The chief end of humankind may not be "salvation," as the Christian tradition defines it. It may be rather to honor, serve, glorify, and celebrate a God whose change of role into a fully cosmic ruler makes him a great deal more awesome and unreadable, stranger, much less like "us," than the tradition permits. Calvinists have always proclaimed a God of that sort, but they have not always believed in what they are describing. "It may be that the task of ethics," Gustafson writes, "is to discern the will of God—a will larger and more comprehensive than an intention for the salvation and well-being of our species, and certainly of individual members of our species. Yet the conditions for this discernment are not such as to provide absolute certainty about God's purposes." Once again, God has grown bigger and less easy to understand.

The English theologian David Edwards makes the withering comment that "most Anglican bishops are content to preach about an omnipotent but fatherly God and are encouraged to do so by their own powerful but benign status in a hierarchy which is also a self-supporting club." The theme of the seductiveness of power again. Many people, Edwards thinks, find it hard to believe in God because they have no imaginative picture of the way he and the world as they know it are related. This may have to do with the fact that the English, like many others, regard religion as a private affair only weakly connected to such public issues as nuclear war and saving the environment. There are "models" of God, obtrusive in the official prayers of the Church, accentuating his absolute dominance over nature, that do not work well at this particular juncture in history. People today need to be reminded of their own obligation to collaborate in the divine work rather than leave everything to an overwhelmingly powerful deity. The English, "although accustomed to a monarchy, deserve to be reminded that any royalist or lordly theology seems nowadays a very odd interpretation of the gospels."

And yet simply deprivatizing religion and cutting God down to size makes for a shallow faith. Edwards is unimpressed with the argument that the core relationship between God and humans must include the whole world rather than being personal, one-on-one. Political and social activists are among those who take this view. Such an outlook, Edwards thinks, is typical of a widespread failure to explore the "depths of personal religion in Christian history." Making our relationship with God a largely public one leads to a sort of despair that such a sovereign being could be present in so lowly and imperfect a place as our own defective world. But how can that world be empty of him when in Catholic and orthodox piety such things of the world as bread, wine, oil, water, pictures, churches, mediate the real presence of God?

The tension between these opposing ideas—how big is God, and where is he?—arises in part out of a sense, at certain times more strongly stated than at others, that for a religion to endure, it must be adamant that God is nothing like us. Even if that means bracketing the doctrine that we are made in his image, a statement open to endless reinterpretation, a religion cannot have staying power unless it posits an unimaginable God. The intensity with which such a belief is held may be buttressed by the contrary impulse, to depict a God who is not quite in the world and not quite beyond it but a little bit of both.

In the age of Newton, a group of thinkers emerged known as the Cambridge Platonists. Their models of the deity were of the "beyond the world" kind. They did not shrink from the idea that God is very big. In fact, their concept of his "immensity" in an immense universe may have inspired Newton's theory of absolute space and time. One of the most engaging of these idealists, rebels against some of the more austere doctrines of the Reformation, was Henry More, who had attended the local school at Grantham where Newton, ten years his junior, was a pupil. More was of a sweet nature. Walking the playing fields of Eton, he later reflected that if he ended up in hell, "where things are full of nothing but cursing and blasphemy," he would be so obedient and uncomplaining God would relent and set him free after a short time.

As fellow and tutor of Christ's, Milton's college at Cambridge, More spent a quiet life studying and teaching, developing his version of a religion as sweet and reasonable as himself. In it he stressed the simplicity of the Christian message and placed a high value on personal experience. Less importance was attached to doctrine, liturgy, outward forms. More's version of "made in God's image" was a belief that the human mind was "a small compendious transcript of the Divine Intellect." At the same time, he was lukewarm to the idea of salvation by scholarship and learning. He thought that Eve committed the unpardonable treason of tasting the forbidden fruit because she was driven by the fear of remaining naive. More could be as anti-intellectual as only a deep-dyed intellectual is capable of being. Drawing nearer to God was the chief end of life, beside which knowledge is "but a vain fluttering, a feather in a man's cap tossed with the wind."

Newton's absolute space, however, smooth, continuous, and endless, presented something of a quandary to More. Where do you put God in that infinite expanse of sameness? Is there one particular spot more suitable than others? Is God nowhere, or everywhere? More got a letter from René Descartes on this matter, stressing that neither God, nor angels, nor the human mind has a definite size and shape; not one of these is spatial. Bang went the medieval idea of God's *being there* in the world. What to put in its place, how to describe exactly how God and the world work together, Descartes could not decide.

More read the letter with some lifting of eyebrows, especially when it came to claiming that nonhuman animals are nothing more than robots, mindless machines. He took issue with the idea that mind is utterly separate and different from body. Rather, mind is a fourth dimension, a dimension of spirit, and God is the fourth, spiritual dimension of the whole vast universe. The multidimensionality of God took care of a point More badly needed to make, as part of his basic theology, that God has an *intimate* connection with our world, while nonetheless not being defined in terms of it. He wrote back to Descartes insisting that God occupies "the whole machine of the world as well as its singular particles." How could God communicate motion to inert matter "if He did not touch the matter of the universe in practically the closest manner, or at least had not touched it at a certain time? Which certainly he would never be able to do if He were not present everywhere and did not occupy all the spaces."

The early makers of the Reformation had not been so precise, so *scientific*, in placing God in such close proximity to his creation. Luther thought God was ubiquitous in a vague sense, present in everything, which meant, contra the Catholics, the priest at mass did not have to transubstantiate the bread and wine into the body and blood of Christ. But Luther veered into paradox when he refuted the inference that God is literally present in mundane things, natural rather than totally supernatural. In an awkward straddle, Luther said that God "exists in every little seed, whole and entire, and yet also in all and above all, and outside all created things." No wonder he conceded that the whole concept is "inscrutable." More, on the other hand, with his belief that theology, as well as science, should be fundamentally simple, did not shrink from essaying a scientifically respectable definition of how God is present in terms of geometry. As the fourth dimension of the cosmos, argued More, God was a mathematical theorem, not a poetic metaphor. Today, William Placher considers that this physical positioning of God, settling him in a specific location, removed God from being the principle behind all things and let him dwindle into a deity who does some things but not others. And as science eliminated mysteries by explaining them in natural terms, there was less and less for him to do. Once he acquired transparency, he was all the easier to discard, to "identify and kill."

It was More's fondness for the plain and simple, and his antipathy to the needlessly complex, that led him to insist on a "transparent" presence of God in the world, and to abandon Luther's useful inconsistency. One of his complaints against the Roman Catholics was that they liked to make the easy difficult. As for the Scholastic philosophers, "What sophistical knots and Nooses, fruitless subtleties and niceties, what gross contradictions and inconsistencies, the Schoolmen and polemical Divines have filled the World with." Science was not seen as knotty in the same way. If you ignored the dense structures of mathematics that underpinned its results, it could be grasped by pretty well everyone. Aaron Lichtenstein, in an impressive study of More, surmises that as a democratic bias crept into the writings of religious thinkers at this time, faith was almost imperceptibly reduced to the lowest common denominator. There was a tacit suspicion that an intellectual religion is neither necessary, nor may it be possible or desirable, for everyone. That attitude was stated with startling frankness by the modern philosopher James McTaggart, who suggested religion may not be digestible by the masses and can safely be left to a privileged few as long as the rest behave themselves decently.

Such a complacent attitude, perhaps only feasible in Oxford and Cambridge senior common rooms, could not survive the brutal conditions of real life and the vicissitudes of human weakness. The writings of the Cambridge Platonists were vigorous and bracing, showing a caliber of intellect that belied their anti-intellectual pose. But their penchant for simplicity, the overriding insistence on morals and an elusive sense of the personal, led to a sadly depleted religious climate in the decades that followed. Something that started as a nostalgic harking back to what was fondly imagined to be a primordial purity ended as a desire to condense religion into a handful of basic principles. One of its weaknesses, in the longer term, was its attempt to situate God as a dimension of ordinary things, an impatience with a theology that did not use plain language in saying whether God is inside or outside his creation.

The drastic attempts in the twentieth century to insist on an absolute difference between the divine and the human may have had its roots in this inclination to make scientifically respectable the idea of the transparency of God, combined with a reluctance to make strict, hard-edged distinctions. Henry More wanted us to be more like God than perhaps we are capable of being. For all his learning and braininess, he wanted to simplify religion and thereby make it easily accessible in its basic tenets. The twentieth-century backlash was vehement in denying that such an enterprise is possible. Karl Barth made the provocative statement that Christianity is essentially strange, outlandish, different. He felt it a waste of time trying to prove Christianity is superior to other products of civilization, or to show that Christian societies are better than non-Christian ones. The temptation to link it to science, to nature or history, to anything whatever apart from the Gospels, is like being mesmerized by a poisonous snake: "It stares back at you, but you hit it and kill it as soon as you see it." Barth was enraged when he heard the words "a natural knowledge of God," which was more or less what Henry More and the Cambridge Platonists had been teaching.

What seems to be called for is a religion that does a tricky balancing act between aspiring to be "like" God—the technical term is *deiformity*—and at the same time being aware that God is so radically unlike us we are undertaking a project that can never be completed. There is even a suggestion that religion ought not to be too healthy-minded, which makes it quite foreign to the spirit of science, the last word in healthy-mindedness. At the end of the nineteenth century, the Anglican Church had grown so used to making religion compatible with the latest discoveries in science and the nonreligious culture that it suffered from severe inhibitions about speaking out against trends in the world clearly injurious to faith. That was the reverse of the predicament of the Roman Catholic Church, which voiced strong objections to what it saw as immorality and worldliness all around, but at the same time refused to have any dealings with progressive ideas in science and the secular culture that a growing number of people had come to regard as a wholesome feature of the onward march of civilization.

A famous convert from Anglicanism to Catholicism at that time looked at the alternatives open to religion and decided that it had no business trying to be as healthy, as sane, as normal, and as optimistic as the prospects for a utopia science seemed to be opening up. That would render God so ordinary, so wedded to a future of comfortable progress, that it was preferable to opt for something less sunny and well adjusted, less benevolent and generously open to new ideas. John Henry Newman, then a cardinal, in a memorable sermon told his audience that England would be better "were it vastly more superstitious, more bigoted, more gloomy, more fierce in its religion, than at present it shows itself to be."

CHAPTER SIXTEEN

Reasoning God into Existence

When Christianity was not doctrine, when it was
just one or two affirmations, but which people
expressed in their lives, God was closer to reality than
when Christianity turned into doctrine. And with
every increase and embellishment, etc, of doctrine, God
became correspondingly more remote. For doctrine
and its dissemination is an increase in appearance, and
God relates Inversely.

—SØREN KIERKEGAARD

T HERE WAS GREAT EXCITEMENT among the Jewish people when
God, nearly seven centuries after his promise to Abraham that he would
make a great nation out of Abraham's descendants, informed the
Israelites that he would conclude a treaty with them as his chosen people. It was three months since they had left Egypt. They were camped in
the desert, in front of Mount Sinai. Calling from the mountain, God
instructed Moses to remind the Elders of the people that they had seen
the marvel of the escape from Egypt, when he "carried you on eagle's
wings." Now the time had come for the Israelites to keep their side of
the covenant. If they were fully obedient, they would be God's treasured
possession, a kingdom of priests and a holy nation.

The Elders were entirely compliant. In unison, they told Moses: "We will do everything the Lord has said." God then announced that in three days he would come down on the mountain in the sight of all the people, but no one was to touch so much as the foot of the mountain, on pain of death by arrows or stones. Only when a ram's horn was blown could they ascend. Meanwhile the people were to consecrate themselves, wash their clothes, and abstain from sex. On the morning of the third day there was thunder and lightning and a dense cloud over the mountain. At the blast of a trumpet, Moses led the Israelites out to meet with God. There was smoke everywhere. The people *and* the mountain trembled.

The Lord seemed worried that the people would come too near. He called Moses to the top of the mountain, and to Moses' apparent impatience, told him to go down and once again warn the people to keep their distance and give him plenty of room. Even the priests must be careful to consecrate themselves, or the Lord would "break out" against them. Moses rather curtly reminds God that he has already told him to put limits around Mount Sinai and set it apart as holy. But God, for the third time, and in almost identical words, repeats that neither the priests nor the people must force their way to come to him. Only Aaron may accompany Moses up the mountain.

This is odd, because the people were trembling with fear in any case, and were not likely to storm Mount Sinai, especially as it was shrouded in smoke. They stayed at a distance and pleaded with Moses *not* to have God come and speak to them. "Speak to us yourself and we will listen," they said. "But if God speaks to us, we will die." Certainly God is under no misapprehension as to the obedience or lack of it on the part of the Israelites. He calls them a stiff-necked people, because they have given him plenty of trouble, at Taberah, at Massah, and at Kobroth Hattaavah. Apparently, the discipline of one Supreme God provokes numerous acts of mutiny, even when Moses is present, even when it is known that the Israelites are the special instrument of God.

On the mountain, God makes it clear that the nation of Israel is to be under his absolute rule and must live by his laws, unlike other nations governed by the laws of kings and secular norms. "Do not make any gods to be alongside me," he warns. "Do not make for yourself gods of silver or gods of gold." As they make their way to the Promised Land, God insists that anyone who stands in their path will have to be driven out and must be made to stay out, in case the Israelites are tempted to worship the obstructing people's alien gods.

Moses is told to approach the Lord, accompanied by Aaron and seventy of the Elders. But only Moses is to come close. The people are to stay where they are. The Israelites accept the terms of the covenant. On the top of the mountain, God receives the delegation standing on a pavement that seemed to be made of sapphire, clear as a cloudless sky. All present eat and drink. After that, Moses spends forty days on the mountain. Only Joshua is with him. He has a word with the Elders before he leaves, however, hinting there might be trouble while he is gone. "Aaron is with you," he says, "and anyone involved in a dispute can take it up with him." For six days a cloud veils the mountain; on the seventh day, God calls to Moses from inside the cloud. He gives Moses the tablets of the law. He also issues minutely detailed instructions for the design of a tabernacle, a "tent of meeting," which is movable and contains an altar presided over by priests. That does not mean, however, that priests can have ready access to the deity. A special part of the tabernacle is to be called the "holy of holies," veiled by a curtain made of blue, purple, and scarlet yarn and finely twisted linen, hung with gold hooks on four posts of acacia wood overlaid with gold. The images of a cherubin must be embroidered on it by a highly skilled craftsman. Only the high priest may enter this special place, and then only once a year.

But while Moses is up on the mountain, receiving instructions, the people down on the desert floor are growing restive. The Elders had been told to refer complaints to Aaron, and soon he was deluged with grievances about the month-long absence of their mediator. Who could say what had become of him? Moses had been promised an angel who would go ahead of the Israelites as they journeyed to the Promised Land, but for now they were left cooling their heels. They ask Aaron to make them a god who will take on that role. Aaron almost jumps to do their bidding. He tells them to take the gold earrings their wives are wearing and fashions them into a golden calf. In front of the calf he builds an altar and announces that on the next day there will be a festival. At the festival, a good deal of eating and drinking and making merry goes on.

God grows very angry at this. He tells Moses to leave him alone so that his wrath can boil up to maximum heat, since he intends to destroy the offenders. Again, Moses has to calm him down. He reminds God of his promise to Abraham, that he would make Abraham's descendants as numerous as the stars in the sky. Then he goes down with the two tablets of the law in his hands, but at the sight of the calf he throws them down, smashing them into pieces. He burns the Golden Calf and grinds it to powder, scattering the powder on the water and ordering the Israelites to drink it.

Why did Aaron submit so easily to the demands of the rowdy mob? When Moses asks for an explanation, Aaron gives the lamest possible answer. He simply states that he was asked to do it, and that he did it, mentioning in passing the gold earrings. "You know how prone these people are to evil," he says to Moses. One way of reading this story is to speculate that Aaron, far from acting at knifepoint, against his will, decided it was necessary to let the Israelites get out of control, to give them an occasion to blow off steam. They needed to relieve the immense stress of living under a theocratic regime where God had become a recluse, threatening to strike dead anyone who came even a little close, while their mediator, Moses, had vanished into the cloud-capped mountain, an absentee prophet negotiating with an invisible God. The historian Norman Cantor thinks Aaron had no option but to let the people run wild, venting their frustrations, and that he did so deliberately. The Israelites, chosen people of a God who would tolerate no competing deities or stand-ins, pioneers of a religion that shunned all the magical props other faiths offered, were a prey to "tremendous psychic pressures." What they had in place of mediators was an excruciatingly meticulous code of laws and prohibitions. By the second century BCE, says Cantor, "the orthodox Jew had no physical or ritualistic crutch. He could not rely on a dying-and-reborn divine savior incarnated in human form, on a material sacrament to unite him with this savior God, on astrological calculation, on touching of sacred objects and other familiar facets of magic. He could rely only on the goodness and majesty of God."

Goodness and majesty, it seems, can suffice only up to a point. In the Book of Exodus, God is keeping his chosen people at more than arm's length; but while they are partying and making obeisance to a golden idol, he is giving Moses the most extraordinarily specific and lengthy blueprint for a tabernacle, richly adorned with gold throughout, no expense spared, a tangible object in which he can be worshipped. In an odd echo of Aaron's request for gold jewelry to make the calf, God tells Moses to solicit offerings of gold, silver, and bronze, blue, purple, and scarlet yarn and fine linen, goat hair, skins dyed red and hides of sea cows, acacia wood, olive oil for the light, spices for the anointing oil and for fragrant incense, and onyx stones and other gems. The bleakness of the Lord's withdrawal and the stringency of his code of conduct is mitigated by the magnificent opulence, the almost excessive splendor, of this one palpable showing forth of his concern and care.

The curious motivations at work in this story suggest that when religion becomes too rational, too tightly laced in by a pedantically exacting set of rules and "Thou shalt nots" given by a God of total sovereignty, it seeks escape in a variation of that religion. For Islam, like Christianity in certain places, the Koran is an unalterable authority, the direct voice of God, a sacred text that contains laws as well as doctrine, an authorized manual of conduct. It makes no distinction between sacred laws and laws for the secular world of politics. Mohammad received it in parts, called *suras*, in a series of revelations when he was forty years old. He did not think he was inventing a new religion, but merely teaching one that had existed since the days of Abraham, an ancestor of Arabs as well as of Jews. Moses was a Muslim, in this view. Divine law is as old as Abraham, and as exacting as that dictated to Moses on Mount Sinai. The word *Islam* means "submission" to God. Muslims are those who submit.

According to tradition, the Koran existed inscribed on stone tablets before the world was made. Its basic truth is that there is no God apart from the one and only God, who needs no partners or helpmeets. In other words, Allah is eternal and irrevocable, untranslatable. By contrast with the Koran, the Hebrew scriptures had been tampered with, Mohammad thought, an accusation that did not ease relations between Muslims and Jews.

In time, Mohammad occupied Mecca and at once set about smashing idols. Like Calvin, he regarded idolatry as the most heinous of sins. He purged the central sanctuary of the city and a muezzin climbed to the roof to sound the Muslim call to prayer. Three years later he led the pilgrimage to Mecca, an ancient pagan tradition which he made into a Muslim one. The destruction of the idols was an acting out of his intense monotheism, his belief in a God who is totally Other, separate from the world he created, though also, as the Koran repeatedly reminds its readers, a God who is merciful and compassionate, and as close to the individual person as "a man is to his own jugular vein."

Does that leave any room for an alternative doctrine? You would think not. But even in this exacting, strict religion of absolute law, some, spurred by need, made an outlet via the otherworldly to a more accessible deity. Within Islam there emerged a mystical movement called Sufism, which, starting in about 800 CE, held that one could achieve union with the unique, transcendent God by cultivating the inner life and practicing asceticism, austerity, humility, and silence. It was a turn to the private from the preeminently public and social dimension of Mohammad's teaching. In time, Sufis hoped to experience moments of ecstasy and, in a numinous merging with the divine, to realize their own annihilation as conscious beings. Early on in the career of Sufism, two Sufis were beheaded for saying they were the same as the deity. One of them, a Persian, cried out at the moment of beheading: "How great is My Majesty."

Later, Sufism was modified to avoid such embarrassing declarations of personal divinity. Abu Hamid al-Ghazali in the eleventh century reduced its intensely private character by emphasizing the importance of doing good to other people, though the movement never lost its doctrine of the possibility of a mystical union with God. The poet Jalal ad-Din Rumi, whose works today crowd the shelves of large chain bookstores in the United States, started the sacred dances now connected with the whirling dervishes, designed to work up a state of religious ecstasy.

There is an important difference between the Koran and the Old and New Testaments. Both claim to have the sanction of the divine. But while most readers of the Bible regard it as the truth filtered through the minds of human authors, Muslims see the Koran as the inerrant, perfect Word of God. The word *Koran* means "the Recitation." It was dictated, not via a person, but direct from the mouth of an angel. Seemingly uncreated, without beginning or end, it existed for all time and before time began. This doctrine is adhered to strictly even today. Suliman Bashear was thrown out of a second-story window by his more devout students at the University of Nablus on the West Bank for suggesting that Islamic religion has a history, that it evolved, instead of appearing complete and mature from the transcript of the angel Gabriel's recitation.

In the history of Islam, the insistence on a severely literal under-standing of the Koran did not go unopposed. One form this resistance took was the Mu'tazilite movement of the eighth century, which took Islam by storm and became the state religion of the Abbasid Caliphate. The word means "Those Who Have Withdrawn." These breakaways wanted a more reasonable doctrine of the relation between God and humans, and a more allegorical reading of the Koran, which would inter-pret seemingly violent and sadistic passages as metaphors for spiritual conflicts instead of hard facts. Ironically, the obligation to think of the Koran as a piece of writing rather than as an eternal, unalterable text was enforced in a highly inhumane manner. Religious leaders were made to say the Koran is not eternal, on pain of being dragged before an inqui-sition. One imam was put in prison and whipped for resisting the new ideology. Here, once again, theology was out of joint with ordinary opin-ion. Lay people hated the idea of taking an oath of this sort, regarding it as a typical product of out-of-touch intellectuals. In a stunning reversal, under a new caliph, it became an offense punishable by death to say that the Koran was created and should be interpreted as well as merely read. An "anti-intellectual rage" swept through Islam in medieval times. At the end of the twelfth century, the religious authorities in Córdoba, Spain, burned hundreds of scientific books as being a disgrace to Islam.

Side by side with this strain of philistinism went a strange sort of vol-untarism. If the Koran must be accepted at the literal level, without any "rational" unpacking of what might be figurative language, then God's intentions are equally out of reach of human interpretation. There is an odd passage in the Koran which curses the Jews for teaching that "God's hands are chained." What a piece of impertinence, Islam considers, for the Jews to say that. God's hands are not chained! They are stretched out, both of them. And he bestows as he will. What the passage means is that God is not limited to the kind of regular, predictable laws that make the universe intelligible to us humans. He is free to do whatever he likes.

There is another echo of Pope Urban VIII and his confrontation with Galileo here. To say that nature works in one particular way and not in any other way is tantamount to saying God *cannot* do what he likes. And once again, too, we see that the limits of worldly knowledge are partly defined by what sort of God a religion decides is the maker of that world. Anyone bent on producing a single, unique explanation for what nature is doing runs up against an immovable doctrine that refuses to accept that a mere scientist can chain God's hands in that fashion. The Islamic scholar Robert Spencer thinks there is potential for serious rupture here. The tension between a religion which can argue that all products of human reason are inferior to divine revelation, and the global hegemony of Western science, he says, "threatens to become explosive." Since it is unlikely that science will change, the Muslim world, to resolve that tension, "will have to reconsider its notion of Allah." This dilemma recalls the crucial debate over whether God is chiefly will or primarily intellect. If the first, his relevance to a human culture, with its modern emphasis on science, the "cool" mediator, tends to diminish.

A vehement expression of dissent to the idea of physics buttressing a theology of divine will and power rather than one of subtlety and intellect was made by the English astronomer Fred Hoyle. An atheist whose wedding ceremony in church to a Christian believer was kept short out of consideration for his nonbelief, Hoyle was an opponent of the Big Bang theory of the origin of the universe. Instead, he preferred a steady-state model in which, rather than the whole universe emerging in a gigantic explosion, atoms materialize one by one, continuously, as the universe expands. New galaxies form from atoms that are perpetually being created, replacing the ones that have moved away. The rate of creation of the new matches exactly the rate of departure of the old. Newton's rule that matter cannot be created or destroyed is replaced by the newer law that while the total amount of matter and energy in the universe remains the same, energy and matter are interchangeable. The fact that atoms can appear out of the "nothing" of energy made the steady-state model at least a theoretical possibility.

Hoyle was quite taken aback by the ferocity of the opposition to his steady-state thesis, actually worked out at Cambridge with the help of two other scientists, Hermann Bondi and Thomas Gold. He was puzzled by the "insensate fury" with which other astronomers tried to demolish it. Could it be, he wondered, that he and his colleagues had unwittingly managed to irritate the objectors on a deeply personal level? Later, Hoyle decided that the real reason for their hostility was that his team was being seen as a menace to beliefs basic to the "theological culture on which western civilisation was founded." The Hoyle group was undercutting the doctrine, central to Christian tradition, of a divinely willed creation of the universe *ex nihilo,* out of nothing. "At first sight one might think the strong anticlerical bias of modern science would be totally at odds with western religion," Hoyle wrote in his book *The Intelligent Universe.* "This is far from being so, however. The big bang theory requires a recent origin of the universe that openly invites the concept of creation, while so-called thermodynamic theories of the origin of life in the organic soup of biology are the contemporary equivalent of the voice in the burning bush and the tablets of Moses."

Michael Buckley, a professor of theology at Boston College and a Jesuit, singles out Thomas Aquinas as largely to blame for the move to discuss God in the intellectual idiom of philosophy, not of religion. Doing so meant stripping away much that dealt with specific religious traditions, customs, unrepeatable experiences. In the prologue to Aquinas's magnum opus, the *Summa,* Aquinas says Jesus showed in himself the way of truth for us. But in the first part itself, Jesus is not included in the five ways, the *quinque viae,* in which God's existence can be proved. Among these are the notion of a prime mover, a necessary first cause, a Supreme Being who is truth, goodness, nobility. From this emerges the classical idea of a perfect being, unalterable, utterly simple and hence indestructible. He is not related to the world, but the world is related to him. Thus Jesus, for Aquinas, is not the primary proof of the existence of God.

During the scientific revolution and after, theologians tended to stick to the Thomistic line. Given the rise of irreligion, they had to find a common framework in which to operate. This shared context was the cosmos. Nature offered evidence that there is a God, and philosophy was the best tool for testing that evidence. Religious *experience* by itself was too private. Any reference to a particular person, which is basic to Christianity, was out. An odd quirk of theology during the period of the Enlightenment was that the life and teachings of Jesus were not widely used to bolster arguments for the existence of God and to rebuff atheism. Thomas Jefferson, practically the personification of the Age of Reason, thought that to do so would weaken rather than strengthen the case for a supreme power. He told John Adams: "Indeed, I think that every Christian sect gives a great handle to atheism by their general dogma that, without a revelation, there would not be sufficient proof of the being of God." Insisting on the uniqueness and supernatural character of Jesus might push waverers into rejecting religion outright. To lure skeptics, it was better to dwell on the wonders of nature and the inner reality of morals, with Christ preeminently a teacher of ethics. Or, why not just let philosophy do the work that theology is supposed to do? There was plenty of talk about Jesus' redeeming acts, but little about the bombshell confirmation he embodied of the reality of God working out his plan in history.

Once more, the disparity between the professional religious thinkers and the ordinary churchgoer was striking. Religion was tangible in the sacred images and paintings, the crucifixes on the walls of taverns and inns, as well as the lavish festivals of saints. Yet the experts were talking of God as if he were chiefly an architect. Science was a willing collaborator in this. "One was informed about God from the outside," Michael Buckley sums up, "as one might be informed about the existence of the New World, or deduce the corpuscular theory of matter."

In this sort of atmosphere there is no "religious density." As time went on, physics, mathematics, medicine asserted their independence from theologies which had relied on them to support their case for the existence of a cosmic God, and these were left bankrupt. Nature took on certain properties once reserved for the divine: "eternal," "infinite." In trying to find a secure base on which to make its supernatural claims, religion brought about its own drift to the margins. One possible strategy would have been to accept the irrational nature of faith; indeed, to glory in it. Some clever thinkers did tread this uncertain road. But the mainstream went in the other direction. They attempted to show that belief in the existence of God is soberly rational and is perfectly compatible with the rationality of science. Many acknowledged that religion contains its own unique and peculiar reality, different from all other kinds of reality, but that was not the view that left its imprint on the religious culture or determined its future.

A secret contradiction arises, Buckley says, when a religion denies its own ability to treat, in an intellectually compelling way, what is fundamental to its nature. Atheism may be the outcome when a theology validates the sacred mainly by converting itself into another kind of knowledge. The very special character of religious knowledge is destroyed in this reduction. A God who is essentially personal must have the personal as the axiom on which the whole theorem of his existence is built. When it is based on an inference from the physical universe, as in the case of Newton (though his excursions into providence and prophecy show that such a theology did not satisfy him), religion defends itself by denying itself, marginalizing its own inherent reason for being, and in time it will become vapid. It is pointless trying to affirm it from the outside, because the same arguments made against atheism will eventually produce atheism, "just as the northern tribes enlisted to defend Rome and its empire eventually occupied the city and swept the empire away."

What happened in the aftermath of the Enlightenment, Buckley predicts, will occur again when two ideas co-exist as internal contradictions: when the culture insists that God is preeminently personal while recruiting as an argument for this claim something quite impersonal, and when religion aspires to prove the existence of the God of religion without any reference to religious experience or personal witness. "To insist that God is personal and to insist as well that human beings have no basic experience or personal manifestation of this personality such as would engage the cognitive claims of religion itself, leaves too great a gulf between experience and reason. In that conflict, reason inevitably gives way to experience."

CHAPTER SEVENTEEN

The Uses of Paradox

Not only did Luther love superlatives; he raised them
to the level of paradox. He loved the paradox—more
than this, it was the life's blood of his theology. There is
nothing astonishing about this statement: it touches the
foundations of Luther's disposition. His love of
paradox is not an overflow from an accidental mood,
not even simply his basic mental and spiritual attitude.
It is part and parcel of the core of his theology, of his
theologia crucis, i.e., of a theology in which contradiction
itself appears as the very sign of truth.

—JOSEPH LORTZ

"This little word 'why' has covered the whole universe
like a flood ever since the first day of creation, madam,
and every minute all nature cries to our Creator: 'Why?'
And for seven thousand years it has received no
answer."

—FYODOR DOSTOYEVSKY

THERE IS NO DENYING THAT certain kinds of mergers between
science and religion can drain religion of its unique character. Part of
that uniqueness is to assert belief in doctrines that defy common sense
and are unprovable. More than one theologian in high standing has pro-

posed that if an article of belief is "absurd," that is all the more reason for accepting it. The English philosopher Roger Scruton suggests that paradox reached its true zenith in Western thought at a time when Christianity began to seize the attention of educated people. St. Paul states plainly that the gospel is "folly to the gentiles and to the Jews a stumbling block." Jesus became weak so that his followers could become strong. Crucifixion, a shameful, scandalous thing, became a means of glory. Christ's lawlessness, his "criminal" status, transcends the law. This is entirely foreign to the thinking of Islam, where it makes no sense at all. Islam has no allegory of the Fall of man, which means there is no need for man to be redeemed by a saviour. It is unthinkable that the one God should come to earth in the form of a human to undergo torture and a hideous lingering death. It would be entirely beneath the majesty of an intensely monotheistic deity to demean himself in that way, and no thinking in terms of paradox could rescue his reputation.

The odd thing about paradox in religion, Scruton thinks, is that it is often an invitation to commitment. There is something in the human makeup that will rush to embrace a statement that boggles belief, to say, "Yes, it *must* be so." The fact that it deals a devastating blow to common sense is part of the attraction. It seems to show the victory of thought over reality. "I can believe anything," it says, "even this. Join me!"

In fact, an important theme in the story of Christian theology is that one way of preserving the special *difference* of religion is to emphasize, even celebrate, the basic irrationality of faith. For the sake of the health of religion, the paradoxes of that faith must not be removed. The doctrine of the Trinity, the account of a god-man submitting to a barbaric execution, the idea that God is good but the world is wicked, that we are made in his image but he is quite unlike us, all are highly contradictory, deeply paradoxical.

Traditionally, a paradox was seen to be most dangerous in the artificial realm of mathematics and logic. When Bertrand Russell was developing the ideas for his *Principia Mathematica,* co-authored with Alfred Whitehead, he came across a paradox so simple it could be written down on the back of a postcard, yet so deadly it threatened to kill his system. Russell could not sleep for many nights afterwards. He decided that common sense had led him into this trap, and he resolved the paradox by throwing common sense overboard.

Formal logic, unlike our informal, real-world thinking, including religious thinking, cannot tolerate a single contradiction, because logic has chosen to sever its connections with the real world and be completely artificial. There is a total absence of context, of a wider milieu in which an inconsistency can be rendered harmless. Once, at a dinner party, Russell remarked that "it is useless to talk about inconsistent things. From an inconsistent proposition you can prove anything you like." Another person at the table objected: "Oh, come on." Russell stood his ground. He asked his challenger to name an inconsistent assertion. "Well," the man replied, "shall we say, two equals one?" Russell accepted, and asked what he should prove. "Prove that you are the Pope," the other man instructed. Russell did not hesitate. "Why," he said, "the Pope and I are two, but two equals one, therefore the Pope and I are one."

The brittleness of logic is defenseless in the face of even a single contradiction. But we in our everyday thinking have a robust ability to use worldly knowledge to put an inconsistency in its place, to smother one piece of bad information with a wealth of good information. The proof of that ability is that we find the story of Russell's dinner party comical. Anyone who lives in the world knows that a pope is not usually an atheistic English mathematician with a reputation for romantic adventures.

Such resourcefulness in the face of contradiction is a recurrent feature of religious faith. A crassly literal reading of the doctrine of the Trinity might be that "three equals one," enabling a logician to invent an even more outlandish assertion than Russell's. But for believers the contradiction is held in suspension while personal experience of the Holy Spirit and faith in the message of the Gospels makes the doctrine "truer than truth." "If I *must* get to heaven by a Syllogism . . . ," John Donne began, in a sermon preached at St. Paul's Cross in 1629. He then proceeded to offer a "logical" proof of the Trinity, which is a travesty of logic, showing that the doctrine makes sense only in the light of revelation, faith, and personal experience.

A key to understanding Tertullian's confusing take on religion and life is the severity of his belief in God's omnipotence. Tertullian felt that God's supreme power puts him beyond the comprehension of us finite humans and yet makes it possible for him to communicate across that great divide. That means he is intelligible *because* he is transcendent, a very odd and contradictory argument indeed. Tertullian made the paradoxical statement that God's enlisting in the human condition as Jesus was "inept," and his resurrection from the dead "absurd." There is no genuine incarnation, according to Tertullian, unless it happens in a fashion that is crazy and impossible. Truth can only be achieved by ineptitude.

Eric Osborn, a scholar of early Christian thought, argues that by widening the context some of the sting can be removed from Tertullian's paradox, just as in the case of Russell's dinner party witticism. The context Osborn points to is the doctrine of God's omnipotence, as stated in scripture. The omnipotent God, who "holds the universe in his hand like a bird's nest," must at all costs remain unique and sovereign. Therefore, if God were joined to humankind in a wholly literal and rational way, his uniqueness would be compromised, as also would be the uniqueness of human beings. Osborn rewrites the paradox: "God is wholly other, and differs from man and from all else. If he is joined to man in a way which is not shameful, inept and impossible, then either God is no longer God or man is no longer man. If God is joined to man in a way which is shameful, inept and impossible, then God is truly God and man is truly man."

But, of course, if you take a firm stand that God is total power, you are at risk of slipping into another incongruity. How can you say, keeping a straight face, that such an omnipotent and perfect being rules over a world whose defects are so blatantly apparent? In the early days of Christianity, Marcion of Sinope, a wealthy shipowner based in Rome, asserted what became a heresy, designed to avoid this dilemma. Marcion, who died when Tertullian was ten years old, held that there were actually two Gods. Rebuffed by the Roman Church, he set up his own religious community with branches all over the empire. His two-Gods theory expounded on certain doctrines of St. Paul and aimed to return to the pure, unalterable truths of the original faith. Marcion wanted to eliminate from the New Testament all but the ten letters of Paul and the Gospel of Luke, the one closest to Paul's ideas. According to him, the world was created by a just and frequently angry God who insisted on strict observance of the law, an injunction often ignored by sinful humankind. Another, superior God then took notice of the plight of mortals and sent his Son to rescue them from perdition. This did not sit well with the inferior, Maker God, who, in the sort of rage that afflicts a person who does not really know what is going on, had Christ crucified. That was not only a cruel and ugly thing to do; it was also quite unreasonable, since Jesus had observed the law to the letter. The inferior God had to admit he was wrong, and handed over to the risen Jesus the souls of those redeemed by his execution.

Marcion appointed himself as Paul's successor with a mission to reform a faith less than a century and a half old. His message was that salvation lies in rejecting the Maker God and acknowledging the superior God and his Son. His theory of radical dualism took care of the problem of why God permits evil in the world, and why matter seems so indifferent and unfriendly. Marcion's book was entitled *Antithesis*. It made a strict distinction between the God of the law and the God of the gospel, the first bad, the second good, thereby aiming to straighten out the contradictions, to simplify rather than make the difficulties more intractable. At the cost, needless to say, of making Christianity a polytheistic religion.

Marcion was perhaps the first but certainly not the last to attempt such dualism. In the twentieth century, Karl Barth set the God of nature at odds with the God of biblical revelation in the most uncompromising terms, desimplifying the whole question for his readers. And while in prison, Dietrich Bonhoeffer felt it necessary to urge his fellow Christians to rediscover the neglected Old Testament, where Marcionites had located the hated Maker deity, a text still seen by some as so alien by comparison with the New Testament that the two might be incommensurable. This was not an attempt to revive the theory of two Gods, but rather to note the existence of a serious tension between the world of nature and human reason and the empire of God. Paul Tillich scolded the Barthians for putting the Saviour God in such opposition to the Creator God that, while they manage to avoid outright heresy, the dualists implicitly "blaspheme" the divine creation by linking it to sin.

Tertullian has attained a certain cult status today, thanks in part to one of his ardent admirers, the novelist John Updike, who has said that writing fiction is a "mediating" function. In Updike's novel *Roger's Version*, there is a sense that two Gods are hovering over the world's baffling career toward who knows what. The story's central character is Roger Lambert, a professor of divinity at an Ivy League college, who is importuned by Dale Kohler, a research assistant for a computer graphics project, to collaborate in a scheme to conjure up the face of God on a computer screen. Kohler is a disheveled, unmade bed of a youth, twentyish and scarred by acne. His extremities are too big for the rest of his body. But on the side he is managing to seduce Lambert's wife. Kohler wants to bypass the paradoxes of divinity by dragging the divine into the open, seized with the idea that the onward march of science has made privacy obsolete.

Roger Lambert, whose favorite theologian is Tertullian, seems, like Updike, to lean to the two-Gods theory. In his mind there is a God who is totally Other, unlike the world and us. This God inspires terror as well as wonder, and he may commit gigantic acts of folly. In this imagining, he turns out to be a monster, full of terrible heat and cold, "breeding maggots out of the dung and so forth," wiping out whole towns with earthquakes, sending plagues and famines. His second deity is the Harvard model, "the one we care about in this divinity school, the living God who moves toward us out of His will and Love." Updike's duotheism may reflect the influence of a childhood spent as a Lutheran but also in the shadow of the Calvinist faith, strong in his hometown.

In an interview with Charles Samuels, Updike said that the God his fictional characters talk about "has some of the hardheartedness that is in the writings of Pascal." He is a God "of the earthquakes, of the volcanos, as much as of the flowers." He suggests there is a double persona, a two-faced Somebody who rules the universe, rather like the subtle distinction in the Old Testament between God and the Lord God, one a little more accessible than the other. "A god who is not God The Creator is not very real to me. So, yes, it certainly *is* God who throws the lightning bolt, and this God is above the nice God, above the God we can worship and empathise with. I guess I'm saying there's a fierce God above the kind God." Is this doubleness? Or is it duplicity?

Dale Kohler, with his acne-pitted immaturity and fools-rush-in bravado, insists there is just a single deity, an architect and craftsman, maker of the universe, a technician working with incredible precision, a *fact*. And facts, sooner or later, must surrender to the relentless conquests of science. "God can't hide any more," Kohler announces, and proceeds to use computer software as a kind of mediator to flush "Him" out of concealment. Kohler is one of those conspicuous unfortunates

met with in Updike's stories, who lose their bearings in the face of paradox. He believes science, in its power of disclosing, has torn away all
God's disguises and protective stratagems. He violates the integrity of
his computer images as he violates the sanctity of Roger's marriage bed.
"They've been scraping away at physical reality all these centuries," he
exclaims, "and now the layer of the little left we don't understand is so
fine God's face is staring right out at us." Roger calls the project blasphemy, a software Tower of Babel.

Kohler hunts God like the *paparazzi* stalking a reclusive celebrity,
barging in on his rights of privacy. His computer program sets up a system of two groups of color blocks, one moving at random, the other
tilted a little toward nonrandomness to introduce an element of intention, of "divine purpose, more or less." During Lent he crashes together
"two conglomerations of vertices and parametric cubic curves," and out
of this shuffle " a face seems to stare, a mournful face. A ghost of a face,
a matter of milliseconds." The face has long hair but no beard. Slowly a
feeling grows that the mysterious presence within the computer is
unfriendly. It seems to hate this intruder, this rash, heavy-handed seeker.
Kohler presses on, hot with the fancy that there will be "an entire body,
or an empty tomb." But the machine turns balky, inimical, bent on shutting Kohler out of its secrets. An orange-red tint smothers the screen. A
fishscale pattern has "a certain optical alignment that leads the eye in,
while yet remaining surface, remaining excited points in a film of phos-
phors backed by a superthin film of aluminum." The search ends in failure: an omnipotent deity with a taste for privacy simply swats away the
most ingenious devices modern technology can muster to force him out
of concealment.

Each in his own way, Dale Kohler and Roger Lambert were trying to
domesticate a divinity that is highly resistant to being tamed. Lambert
wanted a second, gentler, and more personal deity, whose separate identity absolved him from complicity in the evil aspects of the world. The
more austere Creator God manifests the kind of majestic indifference
Galileo ascribed to nature and its laws. At one point Updike bizarrely
puts this divinity on a par with President Ronald Reagan, who "was perfecting his imitation of that Heavenly Presider whose inactivity has held
our loyalty for two millennia." Kohler's method was to treat God as a
scientific object, something like a quark, a subatomic particle which

eludes direct observation but can be unmasked by the brute-force compulsion of a supercollider. The alternative in each case is paradox. A God who is absent but also subtly present in the most ordinary affairs of the world.

This suggests, in the tussle between the Lutheran and Calvinist threads in Updike's religious makeup, that Luther has gained the upper hand. Lutheranism is adept at holding opposite ideas in a balance: we are both sinners and yet saved; closer to genuine faith when in hopeless despondency than when most cocksure; predestined yet possessed of free will. Calvinism, by contrast, lacks the strange and somewhat consoling idea of the world as deeply and richly ambiguous. It holds accurate knowledge in high esteem, whereas Lutherans prefer a certain seat-of-the-pants approach. They can live with hope if certainty is not available. They are two different faiths and it makes no sense to say their Gods are identical.

Luther luxuriated in paradox. This irritated the humanist scholar Erasmus, who liked to keep things simple and thought Luther was undermining our respect for reason. Luther treated with suspicion any theological statement where paradox was missing. "Yes" cohabited with "No." The kingdom of God exists, but people who act with the deliberate intention of finding it will not succeed. Such paradoxes hover on the brink of atheism. Lutheranism, said Updike, "would have a certain otherworldliness, and yet an odd retention of a lot of Catholic forms and a rather rich ambivalence toward the world itself."

Such a strenuous leashing together of incongruous assertions was nowhere so manifest as in "Luther's Version" of what God is. For Protestants, God was a force to be reckoned with rather than a helper always on hand in time of need. He was still a tower of strength, but as Steven Ozment puts it in his masterly account of Protestantism's birth: "His cannons now seemed to roll about loose." In the Lutheran tradition, unlike that of Thomas Aquinas, God's will and dominion took priority over his love and goodness, which nonetheless were still part of his now complex makeup. That made the nature of God a far more urgent question for Protestants than whether one's life was of high moral excellence. The key issue was, will God keep his word and live up to the magnificent endorsements of him contained in scripture? There remained a nagging uncertainty: suppose those excellent opinions are just literary hyperbole? The crux of the matter, Ozment says, was whether the seeming contradictions in God's nature that people experience in their everyday lives are true also in eternity. "The question that drove Martin Luther to near despair was not whether he himself was a sheep or a goat, but whether in the end God was carnivorous or herbivorous. Until the latter question was resolved, the former remained meaningless."

For a question so deceptively simple, there can never be a quick answer. This suited Luther fine, for he was a man who had a contempt for easy answers. His belief that things were more complex than rational fed his disdain for Aristotle, who attempted to explain anything and everything in a cool, reasonable fashion. As a young tutor at the new university at Wittenberg, Luther was required to teach Aristotle's *Ethics*, and the experience filled him with disgust, which later became an "obsession." Luther had a similarly low opinion of Thomas Aquinas. How dare these two presume to unravel something as enigmatic as the universe? Metaphysics, trying to explain sacred mysteries, to remove all contradictions, he said, "is like a cow looking at a new gate." He knew God was beyond all human understanding; but on the other hand he could not stomach a nonchalant treatment of the divine as an abstract principle, which was "Aristotle's Version." He wanted God to be "a person with a name and a will, acting, speaking, giving grace, withholding grace, sustaining the world with his power and directing it till doomsday."

Perhaps there is no more striking example of Luther's gift for paradox than in his treatment of Satan. The voice of this fallen angel sounding in his head articulated Luther's doubts about the character of God, about the goodness of the world, and about his own salvation. On days when this inner demon was in good form, Luther was assailed with qualms as to whether God even existed. To his mentor, Johannes von Staupitz, he once said: "Dear Herr Doctor, Our Lord God does so horribly with people all around. How can we serve him when he smashes people so?" Staupitz, a cultivated man, told Luther he lacked a developed sense of the incongruous; smashing ingrates often does them a world of good. But the doubts did not go away entirely, and Satan loomed larger as Luther grew older. His demonology has been called the rudder that guided his thought. Satan became "a personalized expression of the powers of darkness." And also the embodiment of a profound paradox. If God is as omnipotent, supreme, and unrivaled as Luther asserted, how can Satan loom so large? Looking toward scripture and its inconsistent and perplexing treatment of Satan, Luther began to reconsider Satan's role in making mischief in the world. In Luther's mind, scripture was the authority on the divine and scriptural inconsistencies nourished his tortured sense of paradox. Satan in the end was, incomprehensibly, an instrument of God's will.

At the same time, Luther suffered terribly from the dilemma of predestination. It seemed so contrary to the doctrine of God's love and compassion, especially Jesus' own preference for the company of sinners. Afflicted with the thought that he might be predestined to hell, Luther came to believe the only response to this whole puzzling doctrine is silence, and he urged his followers to ignore the subject altogether. It tended to show Jesus not as a bearer of compassion, but as an agent of merciless retribution. Better to concentrate on the other side of the story, the one that presents Jesus as hurt, humiliated, and condemned. In fact, as his career moved on, Luther tended to avoid thinking too much about what sort of God the Lord of Dominion was. He felt the human mind was simply not up to the task. He took comfort from the idea that this opaque Maker, about whom we cannot say whether he is cruel or kind, sent his Son, and the Son we know is a courier of love. Staupitz had emphasized God's stupendous power, even as he was aware, like other cultivated people, that the world was in a terrible mess. And if God is wholly sovereign, then he must be utterly mysterious, because otherwise there is no rational explanation for why the world is not as perfect as he is.

"Luther could not speak of God's love without thinking of God's wrath; he could not mention Christ without thinking of the dark side of God that lay behind or beyond Christ," Richard Marius, Luther's biographer, sums up. "This was not a rational or systematic answer. It left open a huge, dark hole in Luther's religious consciousness. Luther headed off the baffling inconsistencies of monotheism by widening the context, looking to such Jesus-oriented consolations as the taking of the bread and wine, the ceremony of infant baptism, reading the Gospels to remove the contradictions of systematic theology, and suspending disbelief, as one would do when reading a poem or watching a play: immersion rather than dissection.

Unlike Luther, Calvin was uncomfortable in the presence of paradox. He wanted knowledge that was straightforward, literal, and scientific. He was above all a serious man, trained in the law, who disliked subtlety for the sake of subtlety. He refused to recognize the oddness in the story of a god-man who let himself be mocked and shamed for reasons unclear even today to many well-meaning agnostics. In spite of Tertullian's best efforts. Calvin would rather humbly accept an awkward truth alien to his normal way of thinking than force it into some clever antithesis. Calvin's view of the Bible was not unlike the Muslim accounts of the Koran, that it came from a supernatural source. Calvin felt scripture was authentic because it is validated, not by the Church or the pope but by the Holy Spirit within each reader's heart. Holy scripture just *is* the Holy Spirit. Faith in the Word might tend to promote an inner, private and personal religion, but Calvin insisted on the absolute omnipotence and sovereignty of God.

He was certainly aware of the anomaly of God's unlimited power as opposed to the human individual's freedom of choice. Calvin called this question the "Labyrinth"—meaning a person can wander in its contradictions endlessly and never find an exit. The best he could do was to say that when something evil happens, God assents to it but is not an accessory to it. He could live with that extremely dubious formula in part because of his radically abstract view of the Almighty.

A recurrent theme in Christian thought is that religion should not be easy and obvious, but difficult, strange, undomesticated, not housetrained. Paradox, by taking common sense and turning it upside down, can move religion out of the everyday. Søren Kierkegaard wrote that "The thinker without a paradox is like a lover without feeling; a paltry mediocrity." Kierkegaard took a deliberate decision that if everyone else was busy making things easy for people, he would take it as his task in life to make them difficult. Like Luther, he rejoiced in paradox. On the question of God's remoteness or nearness, that age-old riddle, Kierkegaard said we cannot even begin to tackle it without resorting to contradiction. So great is the majesty of God that the boldest of all devices imagination can invent would miss the point that majesty, which, because it is so *different* from what imagination can conceive, is paradoxical. If he were recognizable straightaway, "he would become ludicrous."

Kierkegaard was not interested in the creeds and doctrines of the

Church. He was struck by the glaring contrast between contented middle-class people who called themselves Christians and the apostles and saints who underwent appalling hardship and suffering for their faith. He wanted to show these complacent and respectable folk that they were not Christians at all in any real sense. This could be done only indirectly and by shock tactics, using arguments that were vastly more oblique than mere logical statements of why it is better to be religious than not to be. True religion, Kierkegaard said, is apt to plunge one into situations where the normal world of ethics and decent morals is turned upside down. Abraham learned that the hard way when he was ordered to sacrifice his only son, a horrifying departure from respectable family values. It requires a huge leap of faith to trust that this terrible decree is somehow part of an overall plan that will make it all come right in the end. But that leap, against everything that is reasonable and normal, is what rescues religion from being trivial and easy and mechanical, and makes it hazardous, even wild and dangerous.

Paradox is also an important feature of modern science. Here the appearance of a paradox is of the greatest interest because it may signal a new discovery, due to the need to break out of an existing framework of ideas. Logic's "brittleness" is actually a spur to innovative thinking. Internal incongruities in modern logic led to the revolutionary proof by Kurt Gödel in 1931 that there will always be true statements that cannot be derived from a given set of axioms, putting final truth out of reach. And it was the paradoxical result of an experiment on the speed of light that eventually led to the stunning discovery of inherent and unresolvable uncertainties in the behavior of particles in the microcosm of matter.

In the late twentieth century, a new phrase was introduced into commentary on modern physics: "ironic science." Its author, John Horgan, who started as a student of literature, noticed that literary criticism today does not purport to get at the hard, literal "truth" about a writer's intentions or the ultimate meaning of a text. Instead, it aims to be "interesting," to throw up arresting insights, ingenious conjectures, outlandish floutings of common sense. Horgan sees in today's science something of the same spirit, a recognition that no final answers may be possible to the deep questions about nature remaining, but plenty of brilliantly inventive speculations. A lay person can hardly contemplate, for example, the idea of a universe of ten dimensions, three of space, one of time, plus six-dimensional Calabi-Yau manifolds located at every point in normal 3-D space, without wondering if it might be all a wonderful flight of the imagination. The counterintuitive elements in today's physics reinforce the idea that God's mind does not work in quite the same way as our minds naturally work. He is not the close intellectual colleague whose thinking meshed so elegantly with that of Newton, who, as Frank Manuel has documented, seems to have considered himself in the possession of a godlike intellect. There is little irony in Newton's celestial mechanics.

The gaps in our knowledge of the physical world may never be definitively closed, given the limits that exist on science's power of discovery. As physics moves forward, it may impose limits on its own scope, and since it is now in a highly advanced, intensely sophisticated stage, the limits are all the more apparent. Three of the most spectacular successes of twentieth-century physics actually created new barriers to understanding, as well as opening fresh insights into nature. These no-entry signs bar the way to ultimate truth. Einstein's theory of relativity prohibits the transfer of information faster than the speed of light. Quantum mechanics rules out the possibility that we can have complete and certain knowledge of the microworld. Chaos theory shows that many of nature's complex happenings are impossible to predict. In view of these prohibitions, we may have to be content in certain cases with ironic science.

Some of today's outstanding questions, John Horgan proposes, could be included among those cases. They pertain, among other matters, to the birth of the universe, the mystery of human consciousness, the origin of life on earth, and whether it would have arisen whatever the quirks of history. Ironic science will probably keep startling us with audacious theories, feeding our appetite for new scientific "revolutions," but it does not make extravagant claims to be the Last Word, the Truth. "At its best," Horgan writes, "ironic science, like great art or philosophy, or, yes, like literary criticism, induces wonder in us; it keeps us in awe before the mystery of the universe. But it cannot achieve its goal of transcending the truth we already have. And it certainly cannot give us—in fact it protects us from—*the Answer*, a truth so potent it quenches our curiosity once and for all time." A canonical statement of ironic science might be that a potent truth is potent for the very reason that, if turned on its head, it constitutes a different but equally significant truth.

Asked to say why modern physics has become so tempting to some theologians, the physicist Cyril Domb gave an answer that would have astonished the virtuosos of the early Royal Society. Domb did not cite the reasons entertained by the admirers of Newton, that science showed the universe as an orderly, stable, and harmonious system. Instead, his reply, brief and to the point, was simply this: "An appreciation of the transience of theories."

CHAPTER EIGHTEEN

Ironic Theology

Irony is an abnormal growth; like the abnormally
enlarged liver of the Strasburg goose, it ends by killing
the individual.
> —SØREN KIERKEGAARD

Deepest down in the heart of piety lurks the mad
caprice which knows that it has itself produced its God.
> —SØREN KIERKEGAARD

I F WE HAVE IRONIC SCIENCE, do we also have ironic theology? Perhaps we do, and if so it has been around for a very long time. Physics has only lately shown that nature may permanently deny prying investigators the privilege of entering its holy of holies, that the barriers to final truth are impassable. But religion has always kept in mind the possibility that some secrets may be forever hidden even from the most devout and sincere believers.

Irony, a blood relative of paradox, often implies a resistance of some kind, a refusal to accept the surface meaning of some statement or text. An ironist says, "No," or, "Not quite," to what is presented as an incontrovertible fact. And a yes may be a no in disguise. Irony dislikes the simple and the apparent. It abhors the naive and does not rule out that the opposite of an accepted platitude may be truer than the platitude itself.

Irony can be an alternative to truth, but it does not insist on that alternative. It leaves all situations open.

There is an element of indifference, of the noncommittal, in irony. A person may say, "Oh I think that's a wonderful idea," to a remark the ironist regards as pure rubbish, but avoids making a constructive response. Sometimes irony is content to let the "rubbish" exist, perhaps feeling superior to it. Irony flourishes in an urbane, sophisticated, bookish culture. It tends to be openly elitist, with a disdain for common sense, clichés, the obvious social niceties.

At times when religious themes are in transit to becoming secular ones, the ironic attitude is a vehicle for enabling a person to take an undefined stance between the two kinds. Daniel Defoe is an example of a writer who pays respect to doctrines such as predestination and divine providence but does not take them *too* seriously. It is as if he had an inkling that under the worldly influences of the Age of Reason, predestination could become "life is unfair" and divine providence the Idea of Progress. Irony can keep a distance from doctrines not to be condemned outright as foolish or wrong, but not to be irrevocably affirmed either.

Some evidence that many people do hold religious doctrines in suspension, at an ironic arm's length, emerged from an opinion survey conducted in England in 2004 by the YouGov polling organization. The survey showed that such an attitude is widespread and intensifying as formal church worship in that country goes into a steep decline. The survey met overwhelming support for the view that the British are now an irreligious people. Only a minority believe God exists and nearly everyone recognizes that Britain is becoming increasingly secular. Yet there is a striking open-mindedness, a readiness to tolerate religious people as well as religious beliefs. The national mood seems to be one of "benign indifference," an ironic stance of noncommittal. Most people regard religion as a consumer good, to be acquired by those who happen to have a taste for it.

More than a third of young British people say they are either agnostics or atheists. But at the same time a respect for tradition and a sense of the "fitness of things" still holds a grip on the popular imagination. A large majority think Queen Elizabeth II should remain as head of the Church of England. "Today's religious doubt frequently amounts to just that: doubt," according to the survey. It hovers between faith and pure secularism. It is at home with the idea that a strictly religious teaching can have a secular translation. And one surprising trend emerges, something quite new: one tenth of the people interviewed who professed a religious belief were *not* monotheists in the usual sense. The irony of a unique Lord of Dominion who rules over the world in which wickedness has a place was sidestepped by the device of alternative deities of varying degrees of perfection. Three percent asserted a belief in more than one God and 10 percent described themselves as adherents of "some other kind of Supreme Being."

It is hardly surprising that in a religious climate as fluid as this England should have produced an ironic theologian, self-described as "dotty," author of a score of books, a television personality preaching a message that encourages religious people to stay unanchored, ever open to new kinds of spirituality. According to the message, one should accommodate the secular, because the sacred *is* the secular.

The theologian in question is Donald Cupitt, a heretic to some, a profound and groundbreaking thinker to others. Cupitt was born in the northern England textile city of Oldham in 1934, the son of an engineer. He was educated at a private boarding school, where he was prepared for confirmation by the Church historian Henry Betterson, a strong proponent of the "argument from design," the doctrine that God's existence can be inferred from the seemingly planned architecture of the universe.

As a student of physics at Cambridge University, Cupitt briefly flirted with evangelical Christianity, with its emphasis on "knowing God." But he backed away when he saw the "psychological tyranny" exerted on members by the leadership. He immersed himself in the Christian mystical writers, especially the Dominican Meister Eckhart, who stated the paradox: "Nothing is so like God as his creatures, and nothing is so unlike." During his third year at Cambridge, Cupitt decided to study for the priesthood and abandoned physics for theology. He declared himself "with qualifications" to be in the tradition of the Cambridge Latitudinarians. In his undergraduate days, Cupitt came to believe that a merciful providence "had bestowed upon the English people the sanest and most reasonable form of Christianity in existence."

Cupitt's first breakaway move from this highly orthodox position was to abandon a belief in the devil so as to protect monotheism "against the enslavement of men by superstitious fear." This contravened the many warnings from clerics, from the Middle Ages onward, that ignoring the devil may begin a downhill slide toward disbelief in the existence of all supernatural beings whatever. Even Defoe, champion of reason in the run-up to the Age of Reason, held that if you believe in God you must believe in Satan. But Cupitt made Satan disappear. And as if to prove those warnings true, as time went on he also made God disappear. In fact, he showed how easy, how "natural," and how thoroughly modern such a vanishing act can be made to seem.

Only one year later, he was arguing in print that a religion cannot be *knowledge* of the divine. Rather, it is a complex way of living. Soon Cupitt, the destroyer of Satan, was being described as a "fallen angel." The hinge event in his apostasy was an onset of doubt as to whether there is any way of describing in human language just what sort of God we are dealing with. Cupitt felt that we can have theories about him, but who is to say that those theories are not like the theories of modern physics, which are "interesting" and have a good fit with observation, but are ultimately futile attempts to describe the indescribable.

Among the formative influences on Cupitt's thought at this time were the writings of Kierkegaard, which he read "surreptitiously," as if dabbling in the exotic and forbidden. His tutor, George Woods, an avowed Latitudinarian, had no high opinion of the Danish thinker. Kierkegaard had been maddeningly evasive on the question of whether God exists as a real, objective being, external to us and therefore something we should at least try to understand. His ironic manner made it unlikely he would ever give a definite, final opinion. There is a striking passage in his *Concluding Unscientific Postscript* in which Kierkegaard seems to say that we ourselves are the conjurers of God's image. "In fables and fairy tales there is a lamp called the wonderful lamp," he wrote. "When it is rubbed, the spirit appears. Jest! But freedom, that is the wonderful lamp. When a person rubs it with ethical passion, God comes into existence for him." In other places, however, a different impression is given. If you are not paying attention, God may rebuke you by ceasing to be a personal divinity, somewhat as he turned his back on Moses.

Cupitt noted with some irritation the prevalence of a clumsy kind of irony in the attitudes of friends and colleagues when talking about religious matters. There was a disconnect between what these people said they adhered to and what they really believed. A strict observance of the

Articles of the Church was in many cases little more than an affectation, Cupitt decided. It amounts to a kind of "religious dandyism," and dandies, as we know, are much inclined to irony. Many of Cupitt's friends who said they believed what the Church required them to believe felt "obliged to make little jokes about it. If misfortune strikes them they say jokingly that God must be annoyed; they joke about thunder and lightning as manifestations of God's displeasure; they joke about praying for rain or for victory in some contest or other, and although they profess in general terms to believe in God's providence and the discernment of God's will, they laugh at the naivete of one who applies these beliefs in a concrete and specific way to just one particular case. This jocularity is significant. It expresses the ironical distance between traditional religious language and the modern autonomous consciousness." And this was the late twentieth century!

When Cupitt was ordained a priest, he came up sharply against this kind of thinking, though without the leavening of sophisticated humor. He chose as his first parish inner-city Salford, a blue-collar town in Lancashire, cradle of the industrial revolution and robustly secular. Here he was required to inform terminally ill patients at the local hospital that their sufferings were due to God's inscrutable will. That experience reinforced his growing doubts about the reality of a being who the Church held responsible for the ugly realities of sickness and pain. In 1968 he wrote a book, *Christ and the Hiddenness of God*, in which he aired his growing confidence that a belief in God is "iconoclastic," that all our ideas of him are flawed, and that when we speak of "God" we are actually talking about a certain kind of human activity, one that seeks to face the dark side of existence squarely, hoping to ease its afflictions.

In 1965, Cupitt was appointed dean of Emmanuel College in Cambridge, but the Latitudinarian in him was fading fast. The original "men of wide swallow" had thought that Newton's science, with its stability, its simplicity, lent support to the idea of God's providential design in nature. During the 1970s, however, Cupitt broke with the idea that science can act as a prop for religion, noting that Newton's argument from design and his literal view of God's providence did not survive the scrutiny of rational minds in the eighteenth century. Cupitt strongly questioned the theory of a "caring" universe. Society does need a cosmology to give religion a public dimension, a lesson he learned from his reading of the Christian mystics, writers he now regarded as too eccentric and introverted to be a guide for moderns. They had no interest in the physical universe. Yet with all its dazzling wealth of discoveries and imaginative flights, today's physics can give us no explanation of the universe. It "systematically and in principle refuses to tell us why we are here, how we are to live, and what we should hope for."

The "fallen angel" had plenty more falling still to do. Increasingly, and more openly, Cupitt was casting doubt on the reality of God. In *Taking Leave of God* (1980), he turned upside down the notion that God "really" shows himself to us, causing a religious awakening. He argued that strictly *religious* considerations may not permit God to be as real as orthodox doctrine expects him to be. In the normal order of things, God comes first, an external and vivid presence, and "religion" second. Cupitt's reversal of this sequence is to say that "holiness," or "spirituality," comes first, and then as part of our struggle to achieve it, we get an idea of what God is. Without the first, there would be no occurrence of the second. Not "religion for the sake of God" but "God for the sake of religion." The inference is that the kind of religious imagining we engage in dictates what sort of God we end up with.

It was an easy step from such a heretical upending of priorities to the conclusion that there is no God out there, which means we must create new religious ways of being. *Taking Leave of God* caused "a hell of a row." Cupitt thought his career in the Church was finished, but he had tenure as a professor at Emmanuel. Robert Runcie, his former tutor, who became archbishop of Canterbury in the same year the book was published, criticized Cupitt in a sermon preached at Little St. Mary's Church in Cambridge, sensationally reported in the local press. In the hullabaloo that ensued, Runcie sent Cupitt a private letter of apology that was later described as "qualified."

Cupitt's heretical trajectory has been from mainstream Anglicanism to a position of extreme ambiguity as to whether God is a real being to a theory that he is an artifice of language, finally arriving at a doctrine of the holiness of ordinary life. In his later work, Cupitt envisaged a new religious epoch in which not only God is made obsolete as a "real" being but also all mediators: gurus, saints, dogmas, priests, the scientific elite, the leaders and teachers of "New Age" spirituality. That puts the whole responsibility for making the universe "mean" something on us humans and our creative powers, which become critically important. In an echo of what happened in the Enlightenment, Cupitt sees our time as one where the "ordinary" assumes a new significance, edging out the transcendent and making way for a greatly more democratic, more horizontal culture. He makes a distinction between what he calls "world" people and "life" people. World people are addicted to grand theories of the cosmos, science, engineering, research. Life people hardly notice the physical milieu and are more attuned to everyday things: art, stories, being happy, loving, reading, talking, looking, making.

Cupitt's friend Iris Murdoch urged him to write novels; the freewheeling nature of fiction she saw as more suited to his treatment of religion as a human construct, not a metaphysical fact. Of theology's strenuous efforts to keep a literal deity, Murdoch said: "The sense of the Divine vanishes in the attempt to preserve it." Later, Cupitt took this idea further. He proposed that the exemplar for "Life" as he defines it is not the novel but the soap opera, a story that is told and retold endlessly, never arriving at a terminus where all things are put right. The narrative is always reinvented, without bringing closure. And a soap opera is *ordinary*; it is about ordinary people finding meaning in the everyday, not in a metaphysical Something outside our world. We have to make our world radiant by the way we live in it. Cupitt by this time was so heretical that his publisher, SCM Press, which had been hospitable to his writings since 1971, turned unfriendly, and he moved to an American house, the Polebridge Press.

Religion, Cupitt decided, has more to do with this thing called Life than with the admittedly splendid discoveries of physics and cosmology. Science is about prizing secrets out of a nature that Galileo had memorably described as indifferent to our curiosity. In Life, there are no secrets needing a priestly caste to wheedle them from a reluctant cosmos. Nothing is hidden. There are no mysteries. Cupitt reads the Sermon on the Mount as calling for a return to ordinariness, an appeal for us to come down from the mists of metaphysics to the realm of the everyday. In the gospel account of the sermon, supposedly preached on a hill near Capernaum, Matthew makes clear that the crowd was deeply impressed, even amazed, because Jesus spoke in a new kind of popular idiom, quite unlike the "teachers of the law" with their empty ritualism and jargon, talking down to the multitude. The sermon throughout concerns itself with human welfare, with family and social relationships, with dealing honestly with others. The phrase "Kingdom of Heaven" is repeated many times, but the way to reach it is by decent and unselfish conduct here on earth. The text is full of ironies: the rich are to be pitied and the meek will inherit the earth. Cupitt began to talk about the "Kingdom of the Ordinary."

In the Kingdom of the Ordinary, the absence of secrets makes the idea of an omniscient God who holds the key to all secrets no longer of much value. Adam was punished for presuming to master all knowledge, to become as omniscient as God himself. But if we are filled with true religion, Cupitt holds, it pervades every corner and crevice of our lives and rules out any "secret compartments or locked doors." Nothing can be withheld from it. In the Bible, it is this very concept of religion that gave rise to the doctrine of God's omniscience in the first place. God's knowledge is always of mysteries and riddles, of things people do not know, cannot know, or do not want to know. But in Cupitt's view, this is exactly what is required of us. What he calls the "religious requirement" insists on bringing to light whatever I have hidden most carefully and kept most deeply buried. In that sense the religious requirement, which is quite different from "God" in the traditional sense, seems to be omniscient, "for it searches the heart and knows me better than I know myself."

There is something rather intimidating about the "religious requirement," however. It imposes a huge responsibility on the individual person, who all on his or her own must ruthlessly expose those parts of the inner self that were tucked away in secret places for our own psychological safety and comfort. Cupitt has dismissed all mediators as superfluous and actually an obstacle to the genuine religious life. This is carrying Protestantism to the ultimate extreme. I need not remind my readers that there have been periods in the history of religion when the removal of traditional mediators has nurtured a peculiarly intense form of anxiety and among certain types of the devout led to the creation of new and different kinds of mediators, some of dubious respectability. Remember Steven Ozment's comment, that the Reformation, by throwing people back on their own spiritual resources, took away much of the familiarity and comfort—the "sweet deceits of traditional piety"—that Catholicism had provided. Such piety accommodates human frailty and folly more generously than the austere Protestant regimen. "Protestant success against medieval religion," Ozment writes, "actually brought new and more terrible superstitions to the surface. By destroying the traditional ritual framework for dealing with daily misfortune and worry, the Reformation left those who could not find solace in its message—and there were many—more anxious than before." In place of the tyrant in the Vatican there was a "new kind of bully," the human conscience.

The demands of the religious requirement may seem more clamorous and insistent than the dictates of mere conscience. There are hints that Cupitt himself needs at least a residue of the comfort tradition and ceremonial can provide. He is still a member of the Church of England. In a singular response to the shabby treatment of a colleague, Anthony Freeman, by the Church hierarchy, he did in disgust hand back to his bishop his license to officiate. But Cupitt says he still "needs" the church community. He retains Anglican orders and takes communion weekly. "He never fully lets go of his love affair with the Anglican Church," says a friend.

One love affair, however, Cupitt did bring to an end: his attachment to the ideas of Kierkegaard. Part of the reason for the breakup was Cupitt's suspicion that the Danish thinker's intense anxiety was unhealthy and spiritually enervating. There is no closure or happy ending in the writings of this spiritual exile but a sense that God can be an adversary. That sobering thought may have occurred to Kierkegaard as a child, since his father, Michael, a wealthy merchant, underwent agonies of trepidation all his life. As a young man herding sheep, racked by poverty and worn down by ill-fortune, Michael cursed the forbidding, judgmental God with whom Moravian preachers in his homeland of Jutland terrified their congregations. The scholar William Hubben says Kierkegaard invites us not to the Kingdom of the Ordinary but to the Kingdom of Anxiety, and to regard unrest, unpeace of mind, and insecurity as the preconditions for becoming truly devout. In the absence of mediators, in solitude, such feelings are all the harder to bear. And Kierkegaard was a notoriously solitary person. He explicitly disowned the eighteenth-century celebration of "the crowd," as a force whose opinions could be aired within the safety of numbers.

In *The New Christian Ethics*, published in 1988, Cupitt affirmed that he had followed Kierkegaard for years, since his "surreptitious" reading of him as a student. But six years later, he announced that such discipleship had been a mistake. He had come to reject Kierkegaard's extreme individualism. The Dane's conception of faith, Cupitt said, was both wrong and destructive and had had a bad effect on him personally. In particular, Cupitt disliked Kierkegaard's "angst-ridden" state of mind, especially in his later writings. In the Kingdom of the Ordinary there should be no deep unease or disquiet, because "Life" as Cupitt envisions it is not deep but horizontal, utterly democratic and inclusive, open, public, accessible. "Being happy" is part of what life is all about. As for God being an enemy, don't worry about it. God has been made to disappear. It is a waste of time trying to conjure him up.

The Church, Cupitt complained, has spent too much time on such a fruitless quest. Theology's attempt to give a recognizable shape to God reminds him of the old *Invisible Man* movies, in which bandages and clothes were wrapped around the Invisible Man to make him visible, and thus like everyone else. Such drapery prevents us from acknowledging that God has no shape, no human attributes. Oddly, however, during the time when Cupitt was most insistent that there is no God to describe, certain scientists, emboldened by the spectacular successes of physics and cosmology, were tentatively suggesting not only humanlike traits for God—"tactful," "considerate"—but also possible entry points for him to intervene in the operations of nature. The notion of a "caring" universe, flat contrary to Cupitt's view that there is a built-in tension between humans and the physical world, made a reappearance in a respectable "scientific" guise.

Cupitt is averse to end points, termini, boundaries, and limits in his ever-shifting account of the religious life. The soap opera of Life is permanently unfinished, a work in progress. It has breadth but not depth. The plot is written anew every day. But in making God disappear, Cupitt does claim to have wrapped up and put away beliefs that have been with us for untold millennia, an act of closure of a very high order of magnitude indeed. Ironically, modern science, open to the strange and the outlandish, to theories that the great physicist Wolfgang Pauli once declared must be "crazy" if they are to stand a chance of being true, is not entirely closed to the strangeness of the divine. Science does not inhabit the Kingdom of the Ordinary. Quite the reverse. That perhaps is part of its attraction for moderns who doubt but do not condemn the doctrines of traditional theism and who are ironically suspended between the sacred and the secular worlds.

CHAPTER NINETEEN

Democratizing Transcendence

So creedless is the American Religion that it needs to
be tracked by particles rather than principles.
—HAROLD BLOOM

For man has not deduced the divine from God, but
rather he has reached God through the divine.
—MIGUEL DE UNAMUNO

THE IDEA THAT RELIGION PRODUCES GOD, rather than God giving
rise to religion, is apt to breed some curious and outlandish offspring. It
can issue in the form of a radically decentered faith summed up in the
slogan: "I decide what God is."

In the age of the Enlightenment, as we saw earlier, the concepts of
"religion" and "the religions" were quite new. The difference between
the two was significant. The first was the general conception of a Christian God as the *pantokrator* of all life; the second was a new understanding of religion in all of its various Protestant guises. It had to do with the
externals of religious life: doctrine, creeds, forms of worship, everything
Kierkegaard reacted against with scorn and derision. All of a sudden
people started talking about religion as if it were a "thing," as if it were
an object to be studied, a chemical substance that could be reduced to
its basic elements. The rise of "the religions" created a confusion and a

crisis of authority within the Christian Church, which prompted inquiring minds to investigate the virtues and defects of competing sects within Christendom and outside it. Without a single infallible Church or concept of God to rule on how and what a person should worship, the need arose for a sorting out of the multiple kinds of observances that were abundant in the wake of the Reformation.

The theologian Peter Harrison, in a study of this important shift in early modern thinking, sees the emergence of a science of religion coinciding with the rise of experimental physics. With the new and powerful methods of scientific research one could compare and contrast different forms of religion with one another. This secular approach was most noticeable in England, thanks to an atmosphere of free expression pretty much without rival in continental Europe. The succession of kings and queens in the period following Henry VIII's break with Rome had presided over a dizzying series of switches from one faith to another. Once, it was thought that England underwent an instantaneous "conversion" from Catholicism to the Protestant faith, like St. Paul on the road to Damascus. Queen Mary's terrible incineration of heretics, according to this view, merely strengthened and confirmed the inevitable embrace of a new religion. The historian Christopher Haigh argues that Reform movements in England were not so much about changing minds as about changing laws, which made possible the changing of minds. "Most of those who lived in Tudor England experienced Reformation as obedience rather than conversion; they obeyed a monarch's new laws rather than swallowed a preacher's new message," Haigh writes. In those circumstances, the "old religion" kept a hold on the affections of large sections of the populace all through the sixteenth century.

Another historian, Christopher Hill, has shown how intense was popular hostility to the established Church in the post-Reformation years. Bishop Thomas Cooper of Winchester spoke of the "loathsome contempt, hatred and disdain that the most part of men" bore to the ministers of the Church of God. And on the part of women, too. Hill quotes one Joan Hoby of Colnbrook in Buckinghamshire, who said of Archbishop Laud in 1634 "that she did not care a pin nor a fart for my Lord's Grace of Canterbury . . . and she did hope that she should live to see him hanged." Laud was indeed put to death a decade later. The years following the execution of Charles I, with the collapse of censorship and the consequent unprecedented liberty of the press, were a time of "glorious flux and intellectual excitement." Ecclesiastical authority and the operation of Church courts had broken down completely. Not until the Restoration of Charles II in 1660 did king and bishops regain control. Fringe religious sects once stifled now congregated openly. They insisted that the choice of ministers must be put to a vote and some argued that there was no point in having a separate clergy. They called for toleration for all forms of Protestantism. And the expansion of commerce broadened English tolerance for exotic faiths. Milton, in his *Areopagitica*, made a comparison between trade and religious tolerance. Free trade went along with an open-minded readiness to understand other ways of worship. One notable import was the arrival of French Huguenots in England, in flight from their persecution after the revocation of the Edict of Nantes in 1685. They set up, mostly in London, as virtually independent communities.

During the Restoration, a revival of strict licensing laws muted anti-clerical voices, though there was an outburst of anti-Trinitarian controversy; but in 1695 the Licensing Act lapsed and was not renewed. A deluge ensued of heretical tracts, verses, satires, and parodies, mostly directed at the Church. Much of this anti-Church sentiment, Margaret Jacob has noted, came from high churchmen.

As the eighteenth century opened, a population explosion was underway in London and other centers of commerce. As we have noted, one result of the expansion of the "public sphere" was a new kind of emergence towards authority. A gregarious public liked to congregate in organized groups: clubs, coffeehouses, debating clubs, places where the social mix was often miscellaneous and the conversation could run free and impudent. The talk in these venues, much of it boisterous gossip and polemical rudeness, held sacred things up to ridicule. The style was often raucous, belittling of authority, and making a defiant claim to superior knowledge. The notion that theologians held the key to all secrets was greeted with hilarity. An anonymous writer in 1705 said that "the humour of being witty grew so much in vogue that to be thought so Men made a jest of the most serious business and even of Religion itself." In a period of contradictions, especially in religion, mockery and wit glibly papered over the deep inconsistencies that centuries of intellectual exertion had failed to resolve.

Sir Richard Blackmore, a physician and poet, spoke out against "Men of little Genius, of mean and poor Design," who "use all their Wit in opposition to Religion and to the Destruction of Virtue and good Manners in the World." Blackmore called on the government to crack down on atheistic writers for the theater. He complained that there were "few strains of Wit or extraordinary Pieces of Raillery, but are either immodest or irreligious." Radical Deists in the Royal Society were another subversive bloc making fun of godly matters. The early Society contained a miscellany of religious sects, including Puritans who were no longer clerics, and converts to Catholicism as well as Anglican Royalists. Some of the leading members, including Robert Boyle and Newton, wanted to retain the idea that scientific work had religious significance. But by the 1820s a group of militant Deists led by Martin Folkes asserted their influence and started an "Infidel Club." Folkes made it a custom in the Society that when any mention was made of Moses, the Deluge, scripture, and the like, it was to be greeted with a loud laugh. It was said that Folkes "Chuses the Councel and Officers out of his junta of sycophants that meet him every night at Rawthmills coffee house."

Religious uncertainty, reeling under the impact of new ideas and expanding horizons, helped to make England more secular. It also led to the comparison of religions and the beginning of a thoroughgoing criticism of the Bible. Up to that time, religion had a sacred history, divinely ordained and guaranteed to be the truth. Now, it also had a natural history. And that meant it had entered the realm of the intelligible.

Once, the idea of "faith" had been a sheet anchor amid the turbulence of competing creeds. An outstanding exponent of that cause was a fifteenth-century German cardinal, Nicholas of Cusa, who made a study of various religions and their history. He decided that a sort of harmony could be discerned among their numerous contradictions by taking into account the limits of human intelligence. Instead of saying that Christians and Muslims have different gods, one could break that impasse by a shift of emphasis: it may be that their mental grasp of the sacred is not the same. After all, the distance between the divine and the human is so vast that any attempt to define God can be no more than an oblique pass at the truth. What goes on between God and the individual is unique and unrepeatable. So of course each person's report of an encounter will not be the same as every other person's. What unifies all of these disparate experiences is faith.

We can be suspicious of the truth of theology, but not of the supreme potency of faith, no matter what shape or guise it assumes. Abraham was the great exemplar of this concept, witness the respect in which he is held by Islam, Judaism, and Christianity. It is the haunting story of Abraham's faith in one God and only one God that makes him such a powerful figure. The horror of sacrificing one's son is so resonant across religious boundaries that all Judeo-Christian faiths find his constancy inspiring. Abraham is named in the daily prayers of Muslims and his story is recited at the opening of the holiest two weeks in Judaism, Rosh Hashanah. It inspires the most sacred hours of Islam's calendar, the Feast of the Sacrifice, and is in the background of Christianity's core celebration, Easter Week. Kierkegaard wrote a whole book about him, citing the episode of God's order to kill his son as showing that religion is a risky and perplexing leap of faith. Bruce Feiler, author of *Abraham, A Journey to the Heart of Three Faiths,* notes that Abraham's ambiguity as an actual character or a composite demands that *we* have faith. He is proof that God is unfathomable, that faith should be all-consuming, and the best example of a new kind of individual emerging at a particular point in the evolution of a religious consciousness. At the time of Abraham, humans had let God down, to the point where he was sorry he had authored the whole enterprise. In the twentieth generation since Adam and Eve—and that includes some pretty flawed Old Testament figures—Abraham appears at last as someone who has real faith. In his story, Abraham is not godlike but quite human, and for that reason needs the strength and direction that only a monotheistic deity supported by faith can provide.

If the Age of Reason produced a split between faith and religion, in our own period another splintering of terms seems to have taken place, this time between "faith" and "spirituality." Some think this betokens a massive shift in religious attitudes that is altering the very nature of religious belief. As many as 12 million Americans are said to be active participants in "alternative spiritual systems," and surveys suggest that although in America most people have some allegiance to their churches and synagogues, their weekday pursuit of "spirituality" veers radically from the teachings of their religious leaders.

A September 2005 cover story, "Spirituality in America," in *Newsweek* magazine asserted that the late twentieth-century "God is Dead" vogue is now itself dead. In its place is a surging interest in the spiritual. What was dying in those years, *Newsweek* wrote, was "a well-meaning but arid theology born of nationalism: a wavering trumpet call for ethical behavior, a search for meaning in a letter to the editor in favor of civil rights. What would be born in its stead, in a cycle of renewal that has played itself out many times since the Temple of Solomon, was a passion for an immediate, transcendent experience of God."

What seems to be emerging is a democratization of transcendence and an abundance of choice as to how to achieve it. "Young people got tired of hearing that once upon a time people experienced God directly," the University of Chicago historian Martin Marty told *Newsweek*. "They want it to happen for themselves. They don't want to hear that Joan of Arc had a vision. They want to have a vision." An opinion poll by Beliefnet, the religious Internet site, showed that nearly 80 percent of people interviewed describe themselves as "spiritual," whereas just over 60 percent say they are "religious." A majority said they practiced religion "to forge a personal relationship with God." Asked if they felt a strong connection with God by studying nature or by praying alone, those who said "praying alone" outnumbered "studying nature" by two to one. Eight out of ten Americans, including 68 percent of evangelicals, believe that more than one faith can be a path to salvation.

An unexpected result of the immigration reform of 1965 was its effect on American religion, the magazine reported. Immigrants brought unfamiliar practices with them when the doors of American hospitality opened wider. Branches of the True Jesus Church from China and the Zairean Kimbangu Church appeared. Beliefnet sends out some 8 million messages of spiritual guidance a day to various clients. There are 460,000 subscribers to the Buddhist thought of the day, 313,000 devotees of the Torah, 268,000 subscribers to Muslim Wisdom, and 236,000 who receive the daily Spiritual Weight Loss message. "Everywhere we looked," *Newsweek* summed up, "there was a flowering of spirituality: in the hollering, swooning, foot-stomping services of the new wave of Pentecostals; in Catholic churches where worshippers pass the small hours of the night alone contemplating the Eucharist, and among Jews who are seeking God in the mystical thickets of Kabbalah. Also, in the rebirth of Pagan religions that look for God in the wonders of the natural world; in Zen and innumerable other threads of Buddhism, and in the efforts of American Muslims to achieve a more God-centered Islam."

Alan Wolfe, a social scientist at Boston College, traveled the length and breadth of the United States to observe new ways in which old certainties are being shattered and then remade. He too found a freewheeling culture of religious experiment, where differences of doctrine no longer matter much and momentous, life-and-death issues of sin and redemption are simply unimportant. Exotic religion is flourishing in America and it is so in tune with the culture that Wolfe finds this familiar, "as strikingly similar to the society in which it flourishes as it is distant from the religion we once knew."

Lest such findings give the impression that unorthodox religious practices betoken an exotic, undomesticated divinity, Wolfe holds that American religion has been tamed. And because we live in an era when religion decides what sort of God it will venerate, God has also been tamed. He is now a friend, rather than the sovereign Lord of traditional belief. With the need for redemption no longer of sobering, tragic moment, we are more at home in the physical universe, a trend some scientists encourage with new theories of a universe peculiarly attuned to life, especially human life. There is not the sense of instability Donald Cupitt stresses. Wolfe calls this "God Lite." He cites a "parachurch" movement, the Aglow International Fellowship, patronized mainly by middle-aged women. This group is inspired by Pentecostalism, which originally regarded it as sinful for women to use makeup and dress attractively. Today, Aglow has turned that rule upside down. Members are encouraged to have manicures and wear smart outfits. To lure new members, it offers a free session at a beauty parlor. Since many are deeply troubled before joining the group, plagued with doubts about their worth, being well groomed helps to raise their self-esteem. Exterior transformation complements internal conversion; but it also expresses the fact that they love God and he loves them, and they must look their best for him.

This may seem a trivial matter. But it turns out to be a great deal more significant than we might think. Looking more attractive is not what one would do for the sake of a distant and remote God. Rather, it testifies to the emergence of a new conception of who God is and what he requires. This new God is a companion who honors the desire of people who *want* to believe in him, "even if I feel mad at him." And God demands less of today's new cohort of believers. The meaning of sin has undergone a drastic alteration because the meaning of God has changed. At a meeting of a Bible Study group in a Presbyterian church in Philadelphia, one of the women members stated in the strongest terms that the image of God as "unknowable but powerful and authoritative" must be thrown overboard. It is almost as if God is going to be known whether he likes it or not.

Wolfe concludes that there has to be a muscular, powerful, and demanding God for a strong sense of sin to flourish. A second indispensable requirement for sin to have any impact on human conduct is a belief in a cunning and potent Satan. As we have seen, the devil figure has haunted the religious consciousness and provided a pious counterbalance to faith for much of the second millennium. He was a thorough nuisance to Martin Luther, especially in the disguise of a good angel, and caused John Bunyan, years after his conversion, to hurl biblical texts at his demonic adversary. Sir Thomas Browne, the eccentric physician-scholar of the seventeenth century, maneuvering in the transitional years when science was gradually clearing the air of superstition, held that Satan was coping with the scientific Enlightenment by the cunning ruse of pretending not to exist. Also, that wicked and subtle imp was putting it about that witches did not exist either, and nonbelief in witches, Browne considered, was tantamount to denying that God exists. Those who doubt witches, he wrote, "do not only deny *them*, but Spirits; and are obliquely and upon consequence a sort not of Infidels, but Atheists."

During the greater part of the century of Milton, Satan remained the most vivid figure in the popular mythology. God had been rationalized through centuries of theology, and was now receding further into the inconceivable as the frontiers of natural causation were pushed back and back. But Satan, the scholar Basil Willey noted, "symbol of how much of the endless indignation of the subconscious against the mind forged manacles of fear, and pride, and rebelliousness—Satan was still far more than an allegory which could be explained in conceptual language." It was probably for this reason that those who resisted the myth-dispelling power of science and held tenaciously to a supernatural outlook "felt that they must cling to Satan in order to keep God."

Today, most English people regard "hell" as simply a metaphor for the excruciating discomfort, disappointment, and wretchedness that punctuates the normal life of the average person. As Jean-Paul Sartre famously said: "Hell is other people." According to interviews conducted with the congregation of the Philadelphia church, Americans seem more likely to believe in heaven than in hell. Many have had supernatural experiences but these, almost without exception, were affirmative rather than forbidding. Any reference to punishment or evil was omitted completely. Some credence was given to angels, but less notice was taken of demons. It will be recalled that in the opinion survey on religious attitudes in England, only about a third of those interviewed believed in heaven, and even fewer believed in hell or the devil. In the popular literature, angels are the embodiment of love and beatific piety; they never scold. There is mention of Satan coming disguised as an angel. In Roman Catholic churches the intimidating emphasis on sin that has historically dominated the dogma is now muted, which is not always to the liking of older churchgoers. Once, it was considered a sin to chew the host at the Eucharist instead of swallowing it. Today, it has been decided that children under the age of seven cannot commit mortal sin and therefore do not need to confess before receiving first communion. "We don't feel holy when we go to church any more," said a middle-aged tool and die worker at a Polish parish church in Detroit. Another spoke for many when he asked: "Who knows what's a sin?"

God, in the fashionable phrase, gives unconditional love. He is just that sort of deity. Many mainstream Catholics feel that God understands "human frailties." The latest Catholic edition of the catechism restates the traditional strict rules, but young Catholics see the classification of sin into categories of lesser and greater as old-fashioned, putting a damper on the positive benefits of belonging to the faith. "America's God has been domesticated," Wolfe writes. "He is there to offer solace and to engage in dialogue with the understanding that, except under the most unusual circumstancs, he will listen and commiserate. In a world governed by this more accessible God, sin still exists and atonement is still possible. But the sins are less numerous, less serious, and more forgiveable."

In bygone days, it might have been a natural thing for the Church to blame a huge natural disaster on the shortcomings of humans, who were being punished for their sins. Of course, it would have been unthinkable in the aftermath of the devastating South Asian tsunami in Christmas week, 2004, to blame the tragedy on the sins of the 175,000 people who died in the floods. In a manner characteristic of these modern times, the reaction of the Anglican Church veered stunningly in the opposite direction. A banner headline in the *Sunday Telegraph* on January 2, 2005, proclaimed: ARCHBISHOP OF CANTERBURY: THIS MAKES ME DOUBT THE EXISTENCE OF GOD. The paper, it turned out, was exaggerating, but only slightly. Dr. Rowan Williams, the archbishop, had written an article in which he stated: "Every single random accidental death is something that should upset a faith bound up with comfort and ready answers. Faced with the paralyzing magnitude of a disaster like this, we naturally feel more deeply outraged—and also more deeply helpless."

He went on to say: "The question, how can you believe in a God who permits suffering on this scale is therefore very much around at the moment, and it would be surprising if it weren't—indeed, it would be wrong if it weren't." Dr. Williams was in fact hovering on the brink of saying that the idea of a God who supervises the orderly workings of a caring universe—Newton's kind of God—is obsolete. Newton's God failed the Church when it became clear that the universe has its own laws and can look after itself. The tsunami catastrophe suggests the universe can behave very badly when it wants to, without heaven's permission. Lambeth Palace, the headquarters of the Anglican Church, issued a denial that the archbishop doubted the existence of God, but the very need of a denial confirmed the troubling implications of the newspaper report.

In eighteenth-century America, sin had its ups and downs, mainly because the prevailing image of God swung violently between the fearsome and the friendly. In midcentury, Joseph Bellamy, a celebrity theologian in New England who gathered a large following after the publication of his anti-liberal book *True Religion Delineated,* took up arms on the hardcore Calvinist side and presented a deity unreservedly omnipotent. Bellamy felt that the tiniest transgression was so heinous it deserves penal measures unthinkably harsh. There is no question of earning a reprieve. Said Bellamy: "a whole eternity of perfect obedience would do just nothing towards making the least amends for the smallest sins." According to Bellamy, human beings deserve eternal damnation because that is the way God wants things to be. By a perverse logic, Bellamy argued that for a person to object to such divine outrages against merely human ethical standards is to diminish God and harbor absurdly grandiose ideas about ourselves. Even sacrificing the life of his Son, Jesus, was not to take away our sin, but to show off God's dislike of human transgression, to give an example of how terrible punishment can be.

Bellamy's tirades demonstrate Calvinism's pragmatic refusal to shrink from the fact of evil or sin. In a study of this phenomenon, the historian Ann Douglas notes, however, that this horrifying doctrine of the immutability of sin possesses tremendous imaginative appeal. She argues that such a conception amplifies the scale of God to mammoth size and unsurpassable majesty: "Crushing, humiliating as it may appear and often was, it could be a source, almost uniquely so even among Western religions, of energy. It provided its adherent, no matter how it belittled him, with a supreme and commanding object of worship."

A man who had long attended Bellamy's church in Connecticut experienced a letdown when Bellamy retired and a new preacher took his place. The replacement, the man said, was satisfactory, but he missed the excitement and stimulation of Bellamy's fire-breathing sermons. "He made God so great—SO GREAT!" the man enthused. Douglas comments: "The terror and the thrill of obeying, partially identifying with, even being punished by such almighty being must have been enormous."

By the mid-nineteenth century, this frightening picture of the deity had changed to a gentler, more parental one. Bellamy's God, the God of Abraham and the Old Testament who had his Son executed to show off his hatred of sin, became one whose love of humanity exceeded his dislike of wrongdoing, who led his creatures into the way of righteousness, by example, not by making them suffer. The assault on the doctrine of God the Hater was led by the Unitarians, who infused a good deal of sentimentality into the story of salvation. God is a wooer, they preached, a conciliator, wanting to save everyone, sinners especially. A loving, caring, and gentle God also implies an image of Jesus that is not the virile, strict Calvinist judge, but a "creature of feeling, passive rather than active, invoking not displaying."

Modern man, said Reinhold Niebuhr, has "an easy conscience." It could be argued that one explanation for a pliable attitude to sin is our gift for inventing the sort of God who has an easygoing tolerance of the wicked inclinations of human beings. And what gives us a license for exercising that gift may be today's cavalier marginalizing of doctrines and creeds.

Today, creeds are given short shrift, if surveys are any guide. Alan Wolfe noticed that many organizers of "small groups," a popular way of giving intimacy to religious discussion, avoid doctrine deliberately. They find it to be divisive and jarring, leading to argument and bad feeling. The aim of the average member of a small group is to "walk with God," not haggle over theological niceties. Those who attend "small groups" say their faith is deepened by the encounters, but they also display a few doctrinal irregularities, believing blithely that Jesus was born in Jerusalem and the Book of Acts is in the Old Testament. Ten percent of Americans think Joan of Arc was Noah's wife! Many evangelicals in the United States regard "religion," with its attendant baggage of sticky and unresolved theological issues, as getting in the way of faith rather than supporting it.

Instead of doctrine, the purveyors of "God Lite" promote feelings. Fundamentalists are supposed to regard doctrine as the unassailable bedrock basis for their movement. They are ready to take on all comers to defend it. They quote scripture in everyday conversation, citing chapter and verse as the final say, unanswerable, on any given topic. For a fundamentalist, even a small difference in biblical interpretation can exile a fellow Christian to outsider status. But such fixation on scripture as the last word, shutting out any further discussion, means that fundamentalists do not need to undertake the kind of rigorous thinking about theology required by other faiths. Since they have found the Truth through faith, what is there left to say? Fundamentalist Christians in the United States are doctrinaire, but they are not particularly interested in doctrine. Evangelicals are more worldly in their outlook and not so fixated on the Word, but they tend to rely on preachers and books to interpret the Bible. Again, the nature of the deity is up for grabs. "These people believe, often passionately, in God," says Wolfe, "even if they cannot tell others all that much about the God in which they believe."

Opinion polls show that more than 90 percent of American Christians believe in God. Ninety percent pray to him on a regular basis and the same number belong to a specific denomination. That is a stunning contrast with the state of religion in England. But it is "spirituality," not conventional piety, that seems to be breathing life into these statistics. A privatizing of religion is taking place, even in the very public context in which it is practiced. The sidelining of doctrine and the indifference to theology are symptoms of this trend. The Swiss theologian Ingolf Dalferth uses the term "cafeteria religion" to describe a flight into spirituality as distinct from faith and the rising interest in the supernatural. In Germany, every seventh person believes in magic and witchcraft; every third person considers the future predictable; and, according to a 1999 survey, there are more seers and fortune-tellers offering their services than all the Protestant and Catholic clergy put together. This allows anyone to put together, from a copious delicatessen of choices, "scraps from the world's religions and natural myths, stress-reducing meditative rituals and esoteric speculations, a pinch of Buddhism and a bit of mysticism after work." Such cafeteria-style worship is carried on under the big tent of the mainstream churches, yet the clergy feels no need to justify any of this in the public sphere.

Theology, says Dalferth, is an odd sort of entity in the present climate of make-it-yourself religion. If people today suppose that the faith they choose from the buffet table decides what or who God is, theology tells them just the opposite, that God determines what faith is and what people are. It is this very concept that may prod churches to be publicly accountable for the fashionable trend toward mysticism, where each person has a private line to God. It may even insist on a public validation of faith. Theology cannot explain why people accept certain things as true, but it can clarify what makes their faith true and rational. This will not happen, Dalferth warns, as long as the churches present themselves as "esoteric cults of mystery and meditation, sing the praises of corporeality and cosmic totality, and hope, through expressive dance, to make up the deficit that they have incurred in relation to private saviours."

Dorothy L. Sayers, who wrote theological works as well as murder mysteries, had much to say about the modern indifference to creeds. She noted that when Jesus rested during a journey through Samaria, he met a Samaritan woman at a well and asked her to give him a drink of water. The woman replied that she was a Samaritan, and Samaritans do not associate with Jews. After a rather sharp exchange, during which Jesus shows he knows more about the woman's private affairs than is comfortable, he tells her: "You worship ye know not what." That suggests, said Sayers with tongue in cheek, that Jesus was "apparently under the impression that it might be desirable, on the whole, to know what one was worshipping. He thus showed himself sadly out of touch with the twentieth century mind."

Our lack of theological rigor is a pity, she went on, because doctrine, in her view, is one of the most exciting things about the Christian religion. She wrote a play, *The Zeal of Thy House*, whose plot revolved around the doctrine of the incarnation and the Trinity. She was surprised to find that these doctrines "were looked upon as astonishing and revolutionary novelties, imported into the faith by the feverish imagination of a playwright." She referred stunned enquirers to the basic creeds and offices of the Church and insisted that her play was thrilling not in spite of the dogma but because of it. The dogma was the drama. Ignorance of Christian doctrine, which she saw as widespread, had led to a perception of God as a big bully, "like a dictator, only larger and more arbitrary," and Jesus as a friendly go-between with no sense of humor who "was meek and mild and preached a simple religion of love and pacifism."

Imagination, not theology, not doctrine, not even reason, is the key to today's turn to "spirituality." Separation of church and state, a bedrock American principle, set Americans free to invent their own deities, rather as breakaway sects did during the newly permissive atmosphere of seventeenth-century England. American Protestantism became a religion of "conscience and decision." Cut loose from the state, the churches lost the hope that with the support of the government they could dictate how society should behave itself. As a result, congregations in a particular church did not duplicate the civil community. Each church was sectarian in its own way. There was a good deal of shopping around and switching from one to another. What mattered was not denomination, but the quality of a person's relation to the God of his or her choice.

No religion is heretical if it can be a medium for spirituality, in the view of millions of today's seekers. The fading of respect for the authority of the Church and its doctrines, in the view of some, is a harking back to that most dangerous heresy that beset the early Christians: Gnosticism. New Agers, of whom there are estimated to be 12 million active in the United States and another 30 million keenly interested, hold that all religions are true in some respects and exist to unite the conscious mind with God. Since the truth is apt to change, it makes no sense to write down a statement of beliefs. Human beings are capable of becoming like God, they think, and a God has no need of an organization to be saved. An absolutely unmediated communion with Ultimate Being is made possible by the fact that the human self already contains a fragment of the divine, the "real me," which has never not existed. Harold Bloom picks Gnosticism as the quintessential American religion because it means being alone with God—no theologians, no priests, just perfect freedom.

As recently as 1986, the Presbyterian theologian Philip Lee thought fit to warn that just as when, in the dawn of Christianity, Gnostic ideas infiltrated the Church, so today "something like the ancient misalliance" is happening in modern Protestantism. Lee brands the new Gnostics as escape artists, dodging all the awkward questions, pretending the dark side of creation does not exist, fleeing from everything except the self. The human tragedy is not a result of sin but of ignorance of God and self. Gnostics cannot abide the idea that a historical, tangible institution like the Church can in any sense be trusted with ultimate truth. The possession of secret knowledge, the whole point of Gnosticism, surpasses the authority of any official organization. Like New Agers, Gnostics think in terms of universal truth, a new heaven and a new earth, a shared, all-encompassing consciousness. God is a totality. That means that the particular, the specific, tends to become unimportant.

In order to rescue modern religion from the one heresy that could push mainline churches to the sidelines, Lee proposes that we firmly assert the claims of what Dorothy Sayers saw as the excitement: the unique, particular, and stunningly paradoxical truths of a religion that hinges on events and human mediation belonging to a specific history as much as to the ages. Churches must make clear the difference between infinite claims and earthly limitations, between privileged knowledge and ordinary faith, the nebulous and the tangible. That may involve doing what Americans of the New Age are reluctant to do: admit that total wisdom is unobtainable in our present existence; that our religious expectations have been too lofty, our efforts to domesticate the eternal overly ambitious and perhaps incompatible with a world that is neither hopelessly bad nor unimprovably perfect, but that could be made better with a little hard work and humility.

CHAPTER TWENTY

God's Biography

What soul ever perished for believing that God
the Father has a beard?

—C.S. LEWIS

Ａs LIBERTINE SPIRITUALISTS CREATE their own form of worship, so too do academics and intellectuals domesticate the divine by subjecting him to the dread discourse of "Bible as Literature." To read God as a literary figure in a narrative where plot twists and turns, surprises are sprung, character evolves, and truth emerges gradually, is to limit the divine to the mere pages of a book. By the logic of literary critics, the ending is the outcome, perfect or imperfect, of the vicissitudes and exertions of all that has gone before. Here ironic criticism merges with ironic theology to produce a deity with a new and complex, vaguely human personality, tamed in the way that centuries of scholarship have rendered the heroes of ancient literature psychologically plausible and therefore less intimidating.

A best-selling type of religious writing today is the "biography of God," in which the Bible is treated as if it were a work by Homer or Shakespeare, where it is legitimate to ask why characters act as they do, what is going on in their minds, and how they are affected by flukes of circumstance and turns of events. Readers of this sort of interpretation are a new breed which, though not fully committed to a belief in God, as specified by the articles of Christian or Jewish faith, like to talk *about* God. Their desire to be religious outside the traditional constraints of religion exactly mirrors the shift in thinking in the pivotal seventeenth century that led to the decay of the transcendent. Like literary criticism, Bible as literature can be exciting and "interesting," without pretending to give us anything like the ultimate truth, the redeeming secret, of faith.

The Bible, in this genre, is not so much the Word of God but rather a literary *oeuvre,* in which God has the leading role. Wade Roof in his *Spiritual Marketplace,* based on his own interviews with a variety of believers and seekers, finds that mainstream American spirituality is "self-consciously metaphoric," meaning that it resists a literal reading of sacred texts in favor of "religious stories," and is strongly imaginative. Readers "understand and appreciate religious narrative as open-ended and generally accept the messiness of life—its struggles, pains and joys—and even try to discern the divine in its midst." God as a literary figure is caught up in this untidy turmoil and he reacts to its pressures and opportunities in a humanlike way. The emerging preeminence of biblical storytelling heralds a clean break between what matters to the ordinary person and what drives the learned wheel-spinning of professional theologians. It also permits one to cavort in the presence of the divine while avoiding any entanglement with the rules and prohibitions of traditional Church doctrine.

One of the most successful of these treatments is *God, A Biography,* by Jack Miles. Born in Chicago, Miles was a Jesuit seminarian for ten years until 1970, studying at the Pontifical Gregorian University in Rome and at the Hebrew University in Jerusalem. Fluent in several languages, he completed a doctorate in the Department of Near Eastern Languages at Harvard. He then embarked on a literary career, acting successively as an editor at Doubleday, executive editor of the University of California Press, and book editor at the *Los Angeles Times.* In *God, A Biography,* Miles starts out with the premise that theology has become a difficult and alien resource for the disciplined few, whereas a book that writes of God as the hero of a literary masterwork is accessible to believers and unbelievers alike. It neither requires nor rules out belief, but rather is a mediator of knowledge, and in that sense enables the reader to enter the "presence" of God. Religion may in this way be thought of as literature that has succeeded beyond any author's wildest dreams.

Miles writes of God's "anxiety"—that word again—due to the fact that he is not a single personality, as one would expect from monotheism, but an amalgam of several. That not all of these personae are compatible makes him more interesting, more "addictive" than a simple deity would be. "God is no saint, strange to say," says Miles, and he sets out with that caveat, as if affixing a consumer warning label on an unsafe product. Much that the Bible says about God is rarely preached from the pulpit because, looked at too closely, it becomes a scandal. Right at the start, soon after the Creation, Miles detects a note of disquiet. God sees that what he has made is "very good." But the narrator does not say specifically that humankind is seen as very good. There are hints of a tension building, an acknowledgment that, as Donald Cupitt likes to emphasize, there is something profoundly amiss, a fracture between the divine and the human which sentimental effusions about seeing God in nature overlook.

Adam is not "very good" until he has a fitting partner. God suggests an array of animal species, "all the wild beasts and birds of the sky," as possible helpmeets, which Adam reject. God then goes to the extreme of actually creating a new being, a woman, from one of Adam's ribs. At first, God tells his human creations to fill the earth and master it. But that does not turn out to be God's intention. Instead, the two are confined to a garden, east of Eden. Adam and Eve have the run of the place, but they are not to eat the fruit of the tree of good and evil. They do, after a cunning serpent tricks the credulous Eve and they decline from immortal to mortal beings. Theology interprets this event as a spiritual Fall, a symbolic, not an actual death. But Miles points out that the narrator of this part of the Bible shies away from the merely symbolic and is fond of stories of deception and fraud. It makes more sense, if we are treating scripture as literature, to read back the discord between good and evil into the mind of the protagonist, the Creator.

Since the serpent is also God's creation, we cannot see the snake as a rival deity or the "fearful watery dragon" of Creation myths in ancient Mesopotamia, its serpentine coils curling like the looping of a great river. Admonishing the serpent, God in effect is censuring himself. "What polytheism would allow to be externally directed anger against a rival deity, monotheism—even a monotheism speaking occasionally in the first person plural—must turn into the Lord God's inwardly directed regret," Miles writes. "The appearance of divine regret, the first of its many appearances, is the first appearance of the deity as a true literary character as distinct from a mythic force or a mere meaning endowed with an allegorical voice. The peculiar, culturally determined interior life of Western man begins with a creator's regret."

Miles sees two Gods in one in the Genesis story. One is Elohim, "God," the other Yahweh Elohim, "the Lord God." God said, "Let us make man and woman in our image, after our likeness." *Then* he told the couple to fill the earth and master it. God makes a world because he wants humankind, and he wants humankind because he wants an image, a reflection, or perhaps a companion. He seems to be entirely alone—a situation he thinks is not good for his first creature, Adam—which suggests he needs some "other" to reflect back his own image. But the Lord God, in a second account of the Creation, does not say humans were made in his image. It is the Lord God who considers Adam incomplete and fashions Eve from one of his ribs. Yahweh Elohim's attitude toward the primal pair is that they might get above themselves and "become like one of us." Later on, he imposes a limit on the human life span of 120 years. The longer the life, the more it resembles the immortal existence of the Creator.

The Lord God can be unpredictably violent, can destroy as well as create. The Flood is an act of obliteration, a mass annihilation of a wicked world. With this act, God becomes the enemy of humankind. He acts out of regret, a human emotion. His words are full of fury and nihilistic rage. "I will blot out from the earth the men whom I created." Noah will survive, and men, women, and animals will start afresh. It is not the case with God (Elohim), who sees the Deluge as a cleansing force, something the world needs to become whole again. During this episode he is unemotional, unruffled, and looks past the cataclysm of the Flood to a new beginning and a new covenant. Once the waters have receded, Noah steps out onto terra firma and makes a burnt offering to the Lord. The Lord accepts the offering and vows never to doom the earth; but even at this hinge moment he reaffirms his opinion that mankind is rotten from its youth, underlining the fact that, unlike God, the creation of human beings was not good. God, on the other hand, is warmer and more generous. He sends Noah a rainbow as a sign that this sort of thing will never happen again, and blesses him and all his descendants.

There comes a point, however, when the Lord God, in Miles's word, becomes "domesticated." As the story of Genesis unfolds, Abraham becomes the key figure. At the time when Abraham is ordered to sacrifice his favorite son, Isaac, he is entirely at the disposal of the Lord. But somehow, when that episode is over—we can't tell whether Abraham really would have killed Isaac when push came to shove—he slips into the position of being able, in effect, to tell the Lord what to do. Possibly there occurs here a synthesis of gods, merging the personal, helping, small-scale gods of the polytheism of ancient Mesopotamia with the high and mighty one who controls the fate of whole nations and promises to populate the earth with Abraham's descendants. In any case, by the twenty-fourth chapter of Genesis, Abraham is treating the Lord as if he were a private assistant, a tremendous comedown for the divine, though it is presented rather quietly, without drama, almost as if it were a logical development. When Abraham's servant sets out on a humdrum errand to purchase a bride for Isaac from among his own people in the Mesopotamian region which he had left some sixty years earlier, Abraham tells the servant that God "will" send an angel to go in front of him for protection. He promises this without consulting the Lord, which is quite a high-handed way of carrying on. Exaggerating just a bit, Miles says this is like sending the Secretary-General of the United Nations on a merely personal errand.

At this juncture, the erratic deity who could create and then violently destroy what he has created, the one with no lofty opinion of the goodness of humankind, becomes a "friend of the family," Abraham's family, and in the process makes himself more knowable and less apt to make decisions whose motives are unclear. One mark of this domestication is that women in the Genesis narrative are less inhibited about taking up issues directly with the Lord, breaking the rule—which is still in force when meeting Queen Elizabeth II, head of the Church of England—that you do not speak until spoken to. Rebekah, the daughter of one of Abraham's nephews and Isaac's new bride, now pregnant with twins, Esau and Jacob, is bothered by the fact that the unborn pair are struggling with each other in her womb. She goes to "inquire of the Lord" as to what is going on. There are no intermediaries, no angels or visions or disembodied voices. It is like going to the doctor. The Lord speaks to Rebekah directly, telling her that two nations are inside her and "two manner of people shall be separated from thy bowels; and the one people shall be stronger than the other people; and the elder shall serve the younger."

It seems now that when women have fertility issues, the Lord is ready to assist. Once, he would speak exclusively to males and was only interested in reproduction on the very large scale. Now, he acts as gynecologist to Rachel and to Leah, making them able to bear children: in Miles's words, "managing the pregnancies one by one." This raises the question, "is the Lord God kind?" He never claimed kindness as one of his qualities, and kindness is often an incidental byproduct of a grand design. Miles conjectures that this kindness is part of the polytheistic composite that describes the divine. And what is true of one constituent of the fusion must be true of the whole. The move from polytheism to monotheism involves choosing which qualities to include in the one God and which to leave out.

Miles reads the Hebrew scriptures as portraying a God who evolves over time and who, while being the main protagonist of the story, is also in a strange way dependent on what the *human* characters do and think. Surprisingly, there is no purely divine action in the Bible, and God does nothing that does not focus on humans. Unlike the Greek pantheon, where Zeus reigns in the company of other gods, the God of the Hebrew scriptures has no social life and his only way of taking an interest in himself is through mankind. "God is like a novelist who is literally incapable of autobiography or criticism and can only deal with his characters creatively; his only creative tactic with them is direction. He tells them what to do so that they will be what he wants them to be. He is not interested in them in their own right."

The God of the Old Testament could not write his memoirs going back to before the Creation because he has no pre-Genesis past, no material for autobiography. More to the point, he had no one to tell the story to. Such a lack of "rich subjective life" makes it very difficult to probe for motives to his actions, except the one explicit aim of making humankind in his image. Often God himself seems not to know his plans ahead of time; he gets irate when humans misbehave, but it is in the experience of becoming displeased with them that he discovers what does gratify him. It is as if God depends on humans for the advancement of his aims, and what humans want is instrumental in this strange process. Miles writes about "the profound originality of a divine-human pact in which both parties complain endlessly about each other." God is, he adds, "impossible to please," but so too are the Israelites, and he complains endlessly about their complaining.

This high Lord is domesticated by virtue of the fact of a mutual irritability, shared by divine and human temperaments. In the Book of Numbers (chapter 11), the people of Israel had scarcely started out on the journey to the Promised Land when they began to grumble to Moses, and then did more grumbling about their hardships "in the hearing of the Lord." At this the Lord waxes very angry and sends fire which scorches some of the outskirts of the Israelites' camp. Members of every family, standing at the entrance to their tents, then start wailing that at least in Egypt they had free fish, as well as cucumbers, melons, leeks, onions, and garlic. Here in the desert they have to make do with manna, so monotonous a diet it makes them lose their appetite. The Lord is annoyed, and Moses is annoyed with him in return, protesting, "What have I done to deserve this?" and, "If this is how you are going to treat me, put me to death right now." The Lord promises to give the Israelites so much meat it will come out of their nostrils and they will grow to detest it. He sends thousands of quail into the camp, driven from the sea by a mighty wind. But while the meat was in the mouths of the people who had said they were better off in Egypt, "the anger of the Lord burned against the people and he struck them with a severe plague."

So the standard theological line, that God is unknowable, is not fully endorsed by the Hebrew scriptures. Until the Book of Isaiah, it is taken for granted that he can be known, because he speaks and acts and makes plans, some of which have to be altered when they come up against a roadblock. There is a learning curve, according to the Bible. In Isaiah, God appears to know everything, but in Miles's reading this omniscience is a fact he has learned about himself, not something he has always known. Still, it is at this point that God becomes mysterious and inscrutable, though he is omniscient about our most deeply concealed secrets, including what we do not even know about ourselves. The fact that God has total access to the human heart and at the same time is a baffling mystery is a core paradox of Western religion.

Miles seems to suggest that God is not "simple." Instead, his actions spring in part from factors in his own makeup, like any literary character. This characterizing or humanizing of God is what has made Miles's book so popular. When he went on a publicity tour to promote *God, A Biography*, he found that his approach to the Bible story loosened up his audiences to talk about their own ideas about God, ideas that owed little to theology. They were emboldened by his portrait of a multifaceted deity who seems at times to be a prey to uncertainty. Miles asserts that God, at least in the early books of the Old Testament, is apt to regret, to change his mind, to have doubts. His audiences reacted strongly to this message, affirming their own inclination to doubt, while at the same time staying faithful to a religion that deals in certainties. "What struck me about those conversations," Miles reports, "was a note of defiance, the defiant rejection of the widespread assumption that doubt and religion are incompatible." His readers brusquely rebuffed the absolutism of the Church's rule that either you "Take it—belief—or leave it—religion" in favor of a middle road of suspension, where creeds and dogmas are simply placed to one side, while acting as if they are true. As one reader put it: "If I may doubt the practice of medicine from the operating table, if I may doubt the political system from the voting booth, if I may doubt the institution of marriage from the conjugal bed, may I not doubt religion from the pew?"

Such a sentiment suggests that even the idea of a God who is not sure whether his creation is worth the trouble it brings is a more sturdy anchor than the dubious artifices human beings invent to put their lives on a firm footing. The doubts so freely aired by readers were not entirely doubts about religion. They also hinted at an uncertainty about politics, marriage, science, the justice system, the pursuit of affluence, the Idea of Progress, and the unfairness of life, secular variations on what were once sacred themes. Yes, you still vote, go to the doctor, propose to the "right" partner, pay a therapist, jockey for promotion at work, watch sports. These activities are in part an alternative to church worship. But what if a society which has leaned so greatly on these options is now undergoing a crisis of belief in itself? Can it be that society is in trouble, and religion, of a rather noncommittal kind, is emerging as one possible answer? Miles surmises that these Americans have not so much recovered their faith as lost faith in the alternatives. But if religion is as open to question as are the institutions of society, then it is likely to be practiced with the same sort of open-minded suspension of belief that characterizes the flawed but nonetheless enduring routines of secular life.

If belief is put on hold, if American religion is a mirror of American individualism, if the church is radically separated from the state, then it could be steeling itself to imagine new forms of godhood, which is exactly what Miles interprets the Bible as doing. When God is unpredictable, contradictory, it is tempting to invent ingenious theories as to his character, just as an endless succession of critics have tried to read the mind of Hamlet. Ironically, those theories are at least based on a narrative rich with incident, feats of enormous fortitude, disasters, reversals of fortune, larger than life figures whose faults are as glaring as their virtues. In sharp contrast, certain types of theology are merely untethered speculation.

St. Thomas Aquinas, still regarded by the Vatican as *the* philosopher of the Roman Catholic Church, is saying in effect that even the portrayal of God in the Bible must be simply disregarded as a means of describing what he "really" is. There are plenty of images of God around, but we must not mistake the image for the reality. A simple God, in his understanding, means there is no way we can separate out distinct qualities in him. Yet the Bible seems to take the opposing view that God is a complex amalgam of different and conflicting pagan gods in a new synthesis that does not always function seamlessly. Miles covers himself by stressing he is not engaged on a theological but rather a literary project. Yet to write a "biography of God" cannot but provoke others to compose their own biographies, which is exactly what Miles's readers were doing. They felt a sense of liberation in "creating" or "writing" the divine.

The doctrine of God's unknowability is easily overlooked because it seems that hardly anyone has heard of it. Keith Ward, who is an outspoken enemy of all crude literal depictions of the deity, including the old man with a beard painted on the ceiling of the Sistine Chapel, finds that almost no one he meets is even aware of such a doctrine. Being informed of it, people react with astonishment, thinking it bizarre, even refusing to believe a theologian in good standing with the Church could ever have said such a thing. They are especially startled to learn that even the concept of the Trinity does not purport to tell us what God really is, and certainly not that there are three different individuals existing inside him. Ward blames science for the popular fallacy that unless something is literally true it is not true in any sense. When people read in the Bible that God weeps, "likes the smell of a good sacrifice," argues with Abraham, and sits on a throne in the sky, they assume it is all meant to be the literal truth. A powerful antidote to this kind of naive literalism is to note how far back the doctrine of unknowability stretches. The early sixth-century CE writer Dionysius took the idea that God cannot be described to such extremes that he came close to saying there is no such being at all. The eleventh-century Islamic thinker al-Ghazali held that the essential nature of God is beyond our understanding, and Brahman, the ultimate reality in the Indian religious tradition, according to the eighth-century teacher Sankara, is totally without qualities, even the ones of omnipotence and omniscience.

People are taken aback when told of these prohibitions on portraying God because they assume, as Jack Miles's audiences did not, that faith means being certain of one's beliefs, with no infusion or residue of doubt. The odder the beliefs, the more they fly in the face of common sense, the stronger your faith needs to be. But Ward argues that faith is not something added to one's factual knowledge, nor does it increase the store of information. True faith may actually diminish our "religious" certainties as we discover how little we can say about God. A Christian and a Jew, he points out, could sit down and have a fruitful discussion by competing as to which of them was able to show he knows less about God than the other.

By contrast, we might think that Jesus ought to be an obstacle to imagining different Gods. His "biography" contained in the Gospels is more like a documentary than the one of Yahweh in the Old Testament. It is intended as an accurate account of what really happened on definite dates in history and becomes less a case of interpreting Hamlet, more like being informed about Shakespeare, whose career contains mysterious gaps and dubious anecdotes. There are stories of Jesus' childhood, his teenage years, and several blank years in his young manhood, whereas Yahweh, unlike Zeus, has no past, no family, no consort. There is an ordinariness about Jesus, as Donald Cupitt emphasizes, and it led some of his early followers and fellow citizens to temper their admiration for him. In St. John's Gospel there is a remarkable episode (chapter 6) where Jesus is teaching at Capernaum, making the amazing claim that he is the "bread of life," that he has come down from heaven to do the will of God and has the power to confer eternal life on those who believe. At this, some of his listeners grow restless and itch to bring him down a peg. How could this young man, with whose parents they are on familiar terms, make such extravagant claims? "Is this not Jesus, the son of Joseph, whose father and mother we know?" they ask. "How can he now say, 'I come down from heaven'?" They refer to him as "this man." Because of this anomaly, this absence of the superhuman, we are told, many of his disciples turned back and no longer went about with him.

The situation is entirely different from that of the Old Testament. Jesus does not acquire definition by making world-shaping decisions and dramatic interventions. He is what he is from the start. His childhood was precocious and startlingly well informed, but otherwise simplicity, everydayness, and an absence of social or political ambitions are part of the profile. He is at ease with unpretentious people. Unlike the opaquely glorious Yahweh, he does not easily lend himself to being refashioned by his followers. In the Old Testament there are episodes in which God becomes a more domesticated presence, but Jesus is both tangible and imaginable from birth. He says he is from another world, but the New Testament reports that some who once admired him were disenchanted to notice that he seemed so much a part of this one.

Nevertheless, strenuous efforts have been made to reinvent Jesus, to adapt him to prevailing ideas of what a God should be like. This move was helped by a theory, fashionable in the first part of the twentieth century, that the historical content of the Gospels, the narrative of the "ordinary" Jesus, must be marginalized. Karl Barth, who died in 1968, actually applauded when scholars discredited parts of the New Testament as being contrary to the historical record. He felt that the false historical claims showed that Christianity and history were quite independent of each other. Barth, writing in the aftermath of the terrible events of World War I, regarded as hopelessly naive, even self-indulgent, the outlook of liberal Protestantism. Its picture of God was one of an amiable mentor of the West's supposed "progress," and its erroneous notion was that religion is just a matter of "walking with Jesus," vaguely sticking to an ethical path that would make a better life for everyone. Many of the liberal theologians Barth criticized had approved the war at its outset. He ridiculed such misguided thinking and reasserted his vision of God as a being so alien, so unimaginable, he can only be approached via a profound faith in Jesus. And by profound faith he did not mean trying to understand Jesus as a historical figure.

Rudolf Bultmann, a New Testament scholar and friend of Barth's, was also averse to a historical reading of the Bible and had a deep suspicion of liberal theology. He absorbed the theory that the early Christians had themselves redesigned Jesus from someone who announced the coming of a new apocalyptic kingdom and merged the story of his life with elements of Greek myth into a divinity as glorious as you could imagine. Bultmann suggested that the Gospels were more or less works of fiction, attempts at creative biography, and as a result made the figure of Jesus so hazy it was possible to imagine all sorts of other Sons of Man. He exonerated his readers from the task of finding the "historical" Jesus because the physical truth could not be disentangled from the mythic elements. In any case, it was of small importance beside a more vital truth, inherent in the *intention* of the writers of the Gospels, to show that human life is anchored in a presence and a power that cannot be explained in literal, worldly terms.

Bultmann was caught between two forces of God's love and history. Though it was clear which he favored, it was possible to say he was trying to have it both ways. Radicals wanted him to demote the Crucifixion, singled out by Barth as the focal point of the Gospels, as dubious history. Orthodoxy, in the form of the United Lutheran Church, officially disowned the practice of demythologizing the New Testament. All the same, the historian Charlotte Allen, in an excellent summary of this troubled period of Bible studies, credits Bultmann with holding off the search for the historical, rational Jesus for a major part of the twentieth century. Even in America, where scholars were more open to the historicizing of Jesus than those in Europe, Bultmann's ideas were popular among cutting-edge Protestant thinkers.

Curiously, however, in the late 1920s, in another reinvention, Bultmann domesticated Jesus, making him more human, an ethical teacher with a liberal bent. In this new vision, Jesus was no longer a mediator, spanning the immense divide between the divine and the human, as Barth had portrayed him. In Bultmann's book *Jesus and the Word*, Jesus is described as being born in the normal way of Mary and Joseph, growing up to be a rabbi who dispensed with the abstemious regimen of John the Baptist's cult to mix with the humbler elements of society. According to

Bultmann, Jesus likely did not think he was the Son of God, the Messiah, but only a messenger of God. Even these supposedly hard facts are only provisional, however, a "reconstruction" open to further reconstruction. The best one can do might be to present an image of Jesus as seen through the eyes of his followers. You could even put the word "Jesus" in scare quotes.

If there is a divine Barth's Jesus and a humanized Bultmann's Jesus, there must certainly be an American Jesus. Stephen Prothero, head of the Department of Religion at Boston University, has written a book with that title, a history not of Jesus himself but of the images of "Jesus" held dear by Americans. There are Jesus films, Jesus musicals, Jesus bumper stickers, tattoos, and a theme park called "The Holy Land Experience" in California. The American Jesus evolved from an abstract principle into a concrete person and then into a personality, a celebrity, and finally an icon and a trademark. During the seventeenth century, Americans regarded Jesus as so exclusively divine they preferred, more than Europeans, to relate to him through Mary as a mediator. This rather austere American portrait stemmed from prevalent Calvinism with its doctrine of the huge abyss that separates the godhead and the world. Puritanism looked more to the Old Testament God and his treaty with mankind than to a close, loving relationship with Jesus. But in the decades after the American Revolution and the specific guarantee of freedom of religion in the Constitution, faith became a matter of individual choice and different faiths competed for converts.

Thomas Jefferson, who was subjected to bruising charges of atheism during the 1800 election campaign, showed his allegiance to Jesus as ethical teacher rather than as divine being by slicing up copies of the New Testament with a cutthroat razor. He discarded passages that seemed to him myth and pasted the rest onto sheets of paper. What was left of Jesus was a meek, benevolent instructor in morals, free of mysticism and philosophy and other extraneous embellishments. Out went the Trinity, the atonement, angels, the Virgin Birth, bodily resurrection, creeds and rites, as well as the "maniac ravings" of Calvin. Only some 10 percent of the original gospel survived. The rest went onto the White House cutting-room floor.

As the nineteenth century advanced, Americans broke with Calvinism as they had broken with the British crown, and set up a multitude of faiths as rivals to the Anglican Church. Evangelical preachers disdained the intimidating "Calvin Gap" between Creator and creature and worked to shrink it by "making humans more divine and God more human." This equalizing went so far that in 1838 Ralph Waldo Emerson complained of the "noxious exaggeration about the person of Christ." Later he wrote: "You name the good Jesus until I hate the sound of him." A crowded marketplace of denominations in prolific confusion led to a reaction in mid-nineteenth-century times, when preachers cut through all the complexity of theological multiplexing by offering a simple, minimalist religion of Jesus. They made him not only intelligible but lovable. Since many who responded to this new approach were women, congregations and theologians, along with preachers, strove to make Jesus more appealing to women.

Jefferson's excision of the Trinity is emblematic of American ambivalence toward Trinitarian doctrine during the history of Christianity. It has to do with what Stephen Prothero sees as the inability even of adaptable polytheists to worship more than one God at a time. Though Christian Trinitarians affirm the divinity of all three persons, the Father, the Son, and the Holy Spirit, "they seem to focus more of their devotion on one of the three. For the Puritans of the colonies, that person was the Father; for many contemporary Pentecostalists it is the Holy Spirit; for nineteenth century evangelicals it was the Son." And the Son was made over in the light of Victorian ideals of the feminine: love, mercy, meekness, humility. Once Calvinism had been sent packing, alternative images of the divine were acceptable, and in some of these Jesus was depicted as an androgynous infant, or as a maternal figure cradling an infant in his arms. That was succeeded in the early years of the twentieth century by a new and more manly Jesus.

In the cultural frenzy of the baby boom explosion of the 1960s, Jesus became a hero of the counterculture almost in tandem with the fashionable "death of God" theology, again monopolizing one person of the Trinity. Many boomers espoused Buddhism and Zen, but many more adopted Jesus as their guru and became, in Prothero's words, "the praying wing of the Woodstock Nation." The vast edifice of the Christian religion, with its magnificent heritage of Old Testament history, was boiled down to Jesus alone. There were Jesus communes and Jesus coffeehouses. Some people dressed like Jesus, wore their hair long and walked the streets of San Francisco as Christ had walked the dusty roads around the villages of Galilee. In 1971, *Time* magazine ran a Jesus Revolution cover. A portrait of Jesus showed him with pink skin and purple hair against a psychedelic rainbow, a new version of the romanticized renditions of him in Victorian years. Ironically, this Jesus, while very much a protagonist of the hoped-for overthrow of the establishment, was definitely not domesticated in the sense of becoming innocuous. He was being groomed as the outrider for a secular apocalypse.

Through all of this there runs a constant theme: the adaptation of Jesus to American culture. There has been no principled separation between the two. The propriety of any reinvention of the Christian tradition is critically dependent on the extent to which we believe God is distinct from the world, on where to draw a boundary line between what is essential to the faith and what is not essential. In all the vicissitudes and bizarre sideshows of American Christianity, Prothero concludes, no one seems to have drawn that line on the other side of Jesus. Throughout that colorful history, no one has regarded Jesus as inessential. Miracles can go out, the Bible as the Word of God can go out, and even God can expire. But instead of eliminating Jesus, Americans have only made him more durable, more popular, more human, floating free of any single set of religious beliefs, a "common cultural coin," and therefore a reducer of the already flimsy margin between the sacred and the secular. Some Americans do not even regard him as Christian, and he has been used to alter and destabilize Christianity itself. Making him more human has not made him more harmless, but it has enabled multiple interpretations, manifold new portraits of him, as restlessly variable as the culture itself.

CHAPTER TWENTY-ONE

The Bagginess of Nature

We do not imagine God to be lawless. He is a law
unto himself.

—JOHN CALVIN

A God you could prove makes the whole thing
immensely, oh, *un*interesting. Pat. Whatever else God
may be, He shouldn't be pat.

—JOHN UPDIKE

I N A MOMENT OF EXASPERATION, Niels Bohr, the great theorist of
quantum physics and usually the most courteous of men, finally blurted
out a reprimand to Albert Einstein that had been brewing for a long
time. Einstein had ruffled the famous Bohr composure by repeating
once too often his now celebrated assertion that God does not play
at dice with the universe. Throwing politeness to the winds, Bohr
enquired of Einstein: "Who are you to tell God what to do?" Though
understandably angry, Bohr was perhaps reading a bit too much into
Einstein's use of the word "God." In invoking the deity, Einstein often
meant no more than the existence of a rational order in the universe.
Sometimes he overstepped this convention, however, playfully flinging
out that he would be "sorry for the good Lord" if observation failed to
confirm this theory. He even attributed some human traits to the deity,

for example, proclaiming that he is "sophisticated but not malicious." Einstein had talked about the "rapturous amazement" he felt at the harmony of natural law, which reveals an intelligence of such superiority that the thinking of mere mortals pales in comparison.

John Polkinghorne, a physicist turned priest with a prolific pen, is also impatient with this attitude toward the Almighty. He singles out Einstein for censure, not quite for his presumption in delimiting the personal preferences of the Creator but rather for ignoring the excitement, the danger, the awful risks inherent in any human encounter with the divine. It is not that Einstein's God lacks majesty and grandeur, though many may argue that scientific discoveries do not lack either. Science confirms the amazing splendor of the lawful operations of the universe. The trouble is that as a picture of God, the highly disturbing being we meet in the Old Testament narratives, it is too docile, too insipid. All is harmony, nothing is discord in His universe. It has an aristocratic elegance and an almost soothing refinement. But "refined" is about the last word we would think of in connection with the melodrama of divine visitation in scripture. As Einstein acknowledged, the universe can leave a person stricken with a sense of inferiority along with moral turpitude and uncleanness, of convulsion rather than awe. That is how the prophets, notably Isaiah, felt in the presence of the divine. Einstein had "domesticated" what is properly a shocking and highly unnerving ordeal.

Isaac Newton made the universe less strange than it had been: more rational, but stripped of much of its charm and mystique. The universe of the Middle Ages was something else. It was replete with vibrant and splendid intimations of the divine. Its concentric spheres possessed intelligence, or "soul." The huge etherial regions between the moon and the *primum mobile*, perfect in love and knowledge, were filled with "shining superhuman creatures." The human imagination, C. S. Lewis said, has seldom constructed anything as sublime as that medieval cosmos. A person could love it as they loved their home city, taking in it a spontaneous, joyful delight. Its meaning was given, not imposed on it by its admirers. It combined "splendour, sobriety and coherence." You could marvel at its magnificence, but at the same time feel at home and safe inside its sheltering borders.

But Lewis had reservations about the supremely regulated *tidiness* of it all, the sense of a place for everything and everything in its place. "If it had an aesthetic fault," he wrote, "it is perhaps, for us who have known romanticism, a shade too ordered. For all its vast spaces it might in the end afflict us with a kind of claustrophobia. Is there nowhere any vagueness? No undiscovered byways? No twilight? Can we never get out of doors?"

Polkinghorne, too, has doubts about the Newtonian cosmos. He faults it not just because of its strained attempt to find something for God to do but because it is too "tame." It depicts a nature that is beautifully ingenious and lawful, but with no off-road wildness, no ferocity, no baffling refusals to act according to the canons of common sense, no bagginess, no weirdness, no turning the traditional rules on their head. Newton himself, it seems, made excursions into the dubious realms of alchemy in part because he suspected the universe might be more than an artful piece of clockwork and was looking for the twilight. Was he, too, a man who hated metaphysics, suffering from a form of claustrophobia?

In fact, isn't theology today taking a too "indoorsy," too housebroken and sanitized view of its subject? Polkinghorne rebukes both biblical literalists and New Age mythologizers for "domesticating" a universe well beyond a homespun scale. He thinks the story of the Garden of Eden is an attempt to reduce the wildness and strangeness of creation and hedge it within a tidy plot of cultivated land where all the animals have names and all the fauna are congenial, except for one troublesome snake. God is within calling distance. Today's popular notion of the world as a living organism is another case of the human desire to make the universe a homier home than the one astrophysicists can offer. The presiding divinity of this New Age habitation is Gaia, an earth goddess, not a cosmoswide one like Ouranos. Worship of her is worship of the quieter, softer, smaller, and safer elements of existence. Meanwhile, science is busy showing us a universe unimaginably vast in its spaces and the time it has been in existence. It is, in the words of one scholar, "Big and old and dark and cold." Big? It is 70,000 billion billion miles across. Old? It has been around for 12,000 million years. Our galaxy, the Milky Way, contains about 100 billion stars, and there are approximately 100 billion galaxies in the universe. Only a few years ago, it was thought that the universe exploded into existence as the result of a Big Bang, giving it a definite birth and dooming it to a definite death, a onetime accident with no larger context. Today, the theory of inflation proposes that whole galaxies are emerging all the time from tiny points of space that seem empty, and new universes will never cease to be made. "Majesty" and "wonder" are words used to describe this theory.

The Stanford physicist Leonard Sullivan believes we have not a single, unique universe, immense as it is, but rather a virtually infinite collection of "pocket universes," each with its own laws of physics, its own collection of elementary particles, forces, and physical constants. Inflation, the brief phase of rapid expansion that set the stage for the Big Bang, blew up the universe to proportions vastly bigger than anything astronomers can detect even with the most powerful telescopes.

Sullivan is an author of String Theory, in which matter is made up, not of point particles, but rather of idealized vibrating threads of energy. In this view, our cosmos occupies a tiny corner of a gigantic "Landscape," which contains a gazillion mathematical possibilities. Some universes would be similar to our own, but differing slightly. For example, the elementary particles would duplicate ours, but gravity might be a billion times stronger. "Not even the three dimensions of space are sacred," Sullivan writes. "Regions of the Landscape describe worlds of four, five, six and even more dimensions."

In such an unimaginably huge conglomeration of universes—an intriguing scientific echo of the theological doctrine that God is completely free to create any kind of world he chooses—the odds are that at least one universe would be a "caring" one, in the sense that its laws of physics are hospitable to life. "The question 'Why is the universe the way it is,' " Sullivan says, "may be replaced by 'Is there a pocket universe in this vast diversity in which conditions match our own?' " Many physical conditions in our universe are delicately balanced, making life possible. They seem like "fantastically lucky coincidences," leading to the speculation that an intelligent designer has been at work. Sullivan's theory of a "megaverse" in which all possible conditions are realized aims to explain, using the ideas of immensity and diversity, how this could have come about by the laws of probability. No matter how immense are the odds against my ticket winning the lottery jackpot, there *is* a winning ticket.

"Every person's concept of God is too small," says John Templeton, a successful financier who works to promote dialogue between theologians and scientists. Templeton thinks a new kind of modesty has begun to express itself as we recognize the vastness of God's creation and our very small place in the cosmic scheme of things. He calls this "humility theology," which may be a subset of what I have called ironic theology. It suggests that God is more grandly cosmic and less graspable by the minds of "humble" humans situated in an inconspicuous corner of a colossal universe. The American astronomer Owen Gingerich has coined the phrase "principle of mediocrity" to convey the idea that if there are inhabited worlds elsewhere in our galaxy, it may well be that we are no more than average, or mediocre in the scheme of things; our intelligence may be superseded elsewhere in the vast spaces of the cosmos. In the mid-twentieth century, when it was thought the Milky Way was the largest galaxy in the cosmos—it had to be; after all, it was ours—the principle of mediocrity might have saved scientists some puzzlement. Another astronomer, Harlow Shapley, had asked a graduate student to work on the problem of why globular clusters around the nearby Andromeda galaxy appeared only half the size of those in the Milky Way. The principle of mediocrity would have led the student to question the premise that Milky Way globular clusters were in fact larger, thereby saving him a lot of work.

So the imagination of scientists has expanded immensely in our day. What about the imagination of theologians? Here we see a different story. "The scale of theological thinking," Polkinghorne says, "in both space and time, still remains domesticated and anthropocentric. When the theologians speak of the 'world' they usually do not mean the universe but our local planet. When they talk of history, it is mostly the few thousand years of human cultural development that they have in mind. When they talk of the future, it seems to stretch only a few centuries onward."

The central doctrine of humility theology is that God is infinitely greater than anything we can say about him. His creativity exceeds our wildest imaginings. Gingerich adopts a seventeenth-century point of view, a voluntarist stance of which Newton might have approved, though it is nowhere mentioned in the Bible: namely, that to say human beings are unique in the cosmos, positing that we are the exclusive focus of God's care and sustenance, is to cast aspersions on his omnipotence and total freedom. The universe need not have been the way it is. Its design was a free choice on the part of the Creator. If that is so, science cannot simply deduce, on the basis of fixed beliefs and a priori reasoning, that this is the way the world should be. It must look and see, measure and experiment and replicate the results of experiments. A humility in the presence of nature entails a readiness to be surprised.

A bigger universe is a stranger universe, one inclined to make God the Creator seem more alien, farther off. It echoes Calvin's thesis that the human individual is an inferior creature trapped in a cosmos designed by a deity our minds are too ungodly to understand. What is different this time is that some scientists, applauded by some theologians, see reasons to domesticate the universe itself. They would make it a vehicle of intelligent life, breeding minds in profusion. The principle of humility is replaced by the anthropic principle, which says biological life in its lowly as well as in its most highly evolved forms is the whole point and purpose of the cosmos. John Polkinghorne is one of the most adventurous of these theorists. He starts from the premise that, contrary to popular opinion, scientists are often more open to the *strangeness* of the world than certain theologians who, bent on rescuing religion from the maw of science, are fixated on what it might be "reasonable to suppose."

Scientists are more likely to ask: What is there that we have evidence to think *might* be the case? They recognize that the universe is full of surprises. Getting to know it means entertaining notions a lot more daring than a pedestrian mind might suppose. "Our powers of reasonable prevision are pretty myopic," Polkinghorne says. Theologians who speculate on the nature of God are not as close to the spirit of science as New Testament scholars who start with the evidence of "the way things were" when Jesus was teaching and his followers underwent a startling transformation from scared and reluctant witnesses to bold, death-defying carriers of incredible and hugely paradoxical news.

Let's look at it this way, Polkinghorne recommends. The bigness of the modern universe may be crushing to traditional ideas about the stature, the uniqueness and special significance of human beings. We might, as Gingerich says, be mediocre on a cosmic scale, just midgets scurrying about on a middling planet in a parochial galaxy. The universe is vast in space and also in time. But if it were of smaller size, more flattering to our status, life would not have had a sporting chance of getting started. There would be no humans around to appreciate its more congenial dimensions. If life is to evolve, it needs time, lots of time. A too rapid timetable of expansion of the cosmos would not have given biology the leisure it needs to produce *Homo sapiens*. We must face the fact that the myth of the cozy little plantation east of Eden was an attempt on the part of nonhumble inhabitants of a universe full of mystery to show that human beings were the work of a caring creator. "Perhaps the desire to make God into a domestic craftsman is because he is more easily tamed in that way," Polkinghorne comments.

In recent years, the conversation between science and religion has moved from a view of God as an underwriter of nature's intelligibility to a more crucial focus on the God of providence who acts in history. This is not a docile divinity. In a tradition deeply embedded in Christian theology, the God of providence has an active role in the world, even if that world is the whole immense arena of the modern universe. Such a notion does not explain why the world is very wicked; but better that God should be incomprehensible than that he stands idly by letting the universe go its own feckless way.

The conception of the God of providence has clear links to the theology of Newton, whose long shadow falls across the radically different world of today's physics. In both cases, there is a strong emphasis on the core fact of monotheism, on the oneness of the Other, and by inference the deep, primordial unity of nature. And if the world is a unity, made by one master builder, then it must be simple at the most basic level. The ancient scientist-priests of Egypt had the right idea. Ardent believers in one God, they studied the many different phenomena of nature as if they were facets of a single creation. Motions of stars in the heavens were of the same kind as movement of things on earth. There was a First Cause, and therefore everything that happened could be traced back to that cause. By contrast, polytheism was harmful to science because it harbored notions of multiple and contradictory causes in nature, each represented by a false God. Frank Manuel thinks this is the real meaning of Newton's curious mention of ancient idolatry in later editions of his *Opticks*.

Unfortunately for the hope of Newton's supporters that his clockwork universe would prove the existence of an active watchmender and watchwinder God, the principle of simplicity eventually meant the universe could do without the services of that kind of deity. In the age of Napoleon, Pierre-Simon de Laplace showed that the Newtonian universe is stable and does not need adjustment by an outside agent. The small eccentricities of the planetary orbits of Jupiter and Saturn are self-correcting. This was a much simpler explanation than the theological one. Asked by Napoleon where he had left room for God, Laplace famously replied that his system had no need of that hypothesis.

If all is simple, all loose ends tied up, no "strangeness" anywhere, then it is hard to see where God can step in to act in his creation. Newton to a great extent made the universe a tight, coherent system of mechanical forces, but he left uncertainties, not least as to how exactly divine correction works. It is via uncertainties that God can be imagined to enter and act in what otherwise would be a self-sustaining world. Newton never finally made up his mind about gravity, and his religiously inclined followers made haste to take advantage of this.

Modern physics depicts a looser, hazier, more fitful and wayward state of affairs than anything Newton could have imagined. Recall, Pope sang, "*Let Newton be!* and all was light," but since then it has been possible to discern patches of darkness, areas of strangeness, of the "twilight" that C. S. Lewis regretted to see disappear. Dynamical systems that are safely and reassuringly open to prediction and control, that are "tame," are not necessarily the most typical ones in nature. It is clear now that most such systems display an exquisitely poised sensitivity to what impinges on them: they can skitter off into a nearly unlimited number of possible ways of behaving. The parts of such systems have a degree of "vulnerability," making it infeasible to say what path one part will take after contact with another part. Even in the apparently simple case of billiard balls colliding, tiny uncertainties in the angle at which a collision takes place result in large differences in how the ball reacts. In the case of a gas, molecules bump into each other vast numbers of times a second. But after only a handful of collisions, the outcome is so hypersensitive it could be affected by a minuscule change in the gravitational field caused by an extra electron on the other side of the universe. There is no hope of predicting exactly how the system will evolve. We are dealing with what Polkinghorne calls an "instrinsic openness" here, a flexibility and a looseness of a kind unknown to the author of the *Principia*, and yet we are still talking about *classical*, not quantum physics. There is cloud-work as well as clockwork. There are laws, but they do not constrain nature in the same old way. "The universe may not look like an organism, but it looks even less like a machine."

It is this openness, this strangeness, that some theologians think could provide an opening for God to act in the world. Keith Ward uses the phrase "creative imagination" to describe a God who prefers a probabilistic, open universe, not a fully determined, clockwork one. It is lawful, but the laws themselves create the indeterminacy. In quantum mechanics, for instance, although at any given time there is a unique set of values for the state of any microsystem, uncertainty comes in when a measurement is made. A famous example is the "cat paradox," devised by the physicist Erwin Schrödinger, author of one of the most important equations of quantum mechanics. Schrödinger's choice of animal for his famous paradox may have been influenced by the fact that an aunt of his mother's who lived at Leamington Spa in the English Midlands had six (later legend said twenty) angora cats. She also kept an ordinary tomcat named Thomas Becket, after the unfortunate archbishop of Canterbury murdered by henchmen of Henry II in 1170. The name was chosen because of Thomas's habit of going out on nocturnal adventures, returning home in a disheveled, half-dead condition. The quantum state of Schrödinger's cat in a box is that the cat is both dead and alive at the same time. It is only when we lift the lid of the box and an observation is made that the cat is either definitely dead or definitely alive. Inside the closed box, the cat is *really* dead *and* alive. The superposition rule of quantum mechanics decrees it. That is part of the strangeness of quantum mechanics, which rules in the world of the very small: perhaps as strange as anything believed by the alchemists of the early modern period.

But the effects of such strangeness are not limited exclusively to the world of the very small. The rules of quantum mechanics have repercussions at the human scale. An example is the fusion reactions that power the Sun, where protons of hydrogen fuse into heavier particles, releasing nuclear energy. For this to happen in the case of two positively charged protons, they have to overcome the tremendous repulsive force of their electrical charges. If classical, Newtonian rules of mechanics were to apply, this barrier could never be overcome and fusion would not take place. The situation is rescued by the quantum principle: the positions of the protons are ambiguous, and this imprecision enables fusion to occur often enough to keep the Sun shining.

Another kind of indeterminacy arose some time after Laplace made his famous riposte to Napoleon. At that time, Napoleon was thinking of the God of the Gaps, which Laplace had closed up. But in the early twentieth century the French physicist Henri Poincaré looked closely at some of Laplace's arguments for the stability of planetary orbits. Laplace had stated as a fact that the 900-year oscillation in the orbital periods of Jupiter and Saturn, caused by a gravitational effect between them, was stable and would continue the same indefinitely. Poincaré, however, came to the conclusion, much to the dismay of astronomers, that no definite answer can be given. Today it has been shown that the area of stability and instability for certain kinds of celestial mechanics is so elusive that in order to decide whether a planet's orbit is stable or unstable we would need to know its initial conditions with greater accuracy than is allowed by the uncertainty principle of quantum mechanics. So it turns out that knowledge of certain events on the large scale depends in a surprising way on what we can know, and what is unknowable, in the world of the very small.

Keith Ward almost paraphrases Newton in his insistence that God does constantly and continually sustain and guide the universe, and is not an external interference with an otherwise smoothly running machine. But he is very *un*-Newtonian in his embrace of the looser rules of today's microphysics. God acts in the physical world in a particular way, made possible by the breaks in physical causality that are a feature of the quantum realm. Here, however, is a dilemma: due to the limits on knowledge implicit in quantum theory, it may be impossible to be sure whether God has acted or not.

Many blunders have been made by attributing to the will of God events that science is unable to explain at a particular juncture of its history. Scientists have egg on their face if at a later time these "unanswerable" questions turn out to have a fully satisfactory answer. The divinity appealed to here is the notorious God of the Gaps. Today there are religious thinkers who hope to avoid this snare by proposing not a God of the Gaps but a God of built-in uncertainties that are an irremovable part of the *process* of nature. Newton wanted to make nature more familiar by rendering it more intelligible. But for today's quantum physics nature, while wholly tamed in that it can be made to account quite adequately for what happens in the microworld, is not "intelligible" in the seventeenth-century sense of the word. All agree as to how it is to be used, but there is no firm agreement as to how to understand it. Perhaps Newton's misgivings over gravity are the closest equivalent to our debate over the interpretation of quantum theory today.

The Stanford philosopher Nancy Cartwright has made the provocative suggestion that unity and simplicity are an illusory goal for modern science. Cartwright draws a radical distinction between laws about things we can observe and laws regarding the elusive "reality" behind appearances, known only indirectly via the medium of theories. The first kind of laws describe what happens, while the second kind attempt to explain what is going on at a more basic level. It is Cartwright's opinion that in today's physics the first kind of laws do their job of description with reasonable success. But in the case of the deeper sort, which are meant to explain, not merely depict, the greater their explanatory power the less adequate is the description. Really powerful explanatory laws of this kind, she says bluntly, "do not state the truth." In fact, she adds, the falsehood of the deeper laws is an unavoidable result of their greater explanatory power, exactly the reverse of what we might suppose.

People who prefer deep explanations often resort to the ancient doctrine of the unity of nature. They look for a superlaw spanning different domains. But Cartwright is suspicious of such a doctrine. She talks about "fictional unifiers" and gives as an example James Clerk Maxwell's laws showing that light and electricity are a manifestation of a single underlying feature. There is no such feature, but a scientist prefers to work with a fictional unifier rather than trying to make sense of a mass of unconnected analogies. Cartwright insists that there is a critical difference between what is true in nature and how we are to explain it. There does not seem to be a law to cover every case. "God may have written just a few laws and then grown tired," she says. "We do not know whether we are in a tidy universe or an untidy one." Insisting on the unity of nature, she argues, is traditionally a French rather than an English trait. The difference, Cartwright says, "is almost theological." One side "thinks the creator of the universe worked like a French mathematician. But I think that God has the untidy mind of the English."

It is typical of the interplay between science and theology that a different view of nature should entail an altered portrait of its creator. The omnipotent God of the age of Newton is today likely to emerge as a subtle, ingenious, and complex divinity who exploits the element of vagueness, the "bagginess," that modern physics has unveiled. Owen Gingerich thinks the physical laws of a created universe could produce a world fit for life, but the starting conditions do not contain sufficient information for us to predict a specific outcome. Yet a particular product could be obtained via the tiny "blur of uncertainty" in the quantum process without breaking any of the laws of physics. So it is not enough to talk only about design, as Newton did. We must also take into account contingency; what is the case, but need not have been the case. These are the new sorts of gaps into which God could be inserted to produce specific kinds of outcome. Gingerich makes the startling suggestion that the extinction of the dinosaurs—making way for the rise of our small primate ancestors and ultimately of the human species—which is thought to have been caused by a gigantic asteroid colliding with the Earth, might have been the result of divine action. A possible site for the crash has been discovered in Mexico. The dust cloud created by such an impact would have blacked out the skies. For months afterwards the rains "would have been as sour as battery acid." It would be difficult to argue that the universe had been designed from the beginning to include that collision, given the indeterminacy inherent in nature, Gingerich says. "But the theist could nonetheless argue without fear of contradiction from our understanding of physical laws, that God could act in a physically undetectable way to guide the asteroid toward its deadly destination."

The idea of divine action via the loophole of quantum uncertainty is controversial, to say the least. Yet it continues to be explored and debated by serious theologians as well as religiously minded scientists. One leading sponsor of the theory is the theologian Nancey Murphy, whose basic position is that any quantum particle has a range of different possible outcomes and God selects one of them. But her position is open to the objection that a quantum system is indeterminate only until a measurement is made, and that need not happen often or at regular intervals. In the case of the cat paradox, God can act only when someone or some event opens the lid of the box and registers the cat as dead or alive. Does God act in every quantum event? Suppose God acts when micro events are amplified into macro ones, as in the case of nuclear fusion on the Sun? He is tightly restricted to occasions when a promising chain of events is available. Divine intervention must submit to the vagaries of nature, which is not how the Old Testament defines the character of providence. The philosopher Nicholas Saunders has taken on the rather easy task of dismantling the theory of a God-guided asteroid wiping out the dinosaurs, pointing out that just to steer into our planet an asteroid which chanced to be near it anyway, by means of quantum tinkering, would take about 3 million years to accomplish if the laws of physics are not to be broken. And if God did act on a regular basis at this level, there would be few quantum processes that would escape his control. In that case, why create the world with quantum indeterminacy in the first place?

John Polkinghorne is not a champion of the new quantum providence. He "dislikes the idea of God scrabbling around in the basement of subatomic particles, furtively adjusting quarks when no one is looking." Of more interest to him is the looseness of nature as revealed by the more recent branch of science called chaos theory. This theory is based on the fact that very simple equations can produce highly unpredictable behavior in a dynamical system on a macro scale. An almost laughably small uncertainty at the start of a process can lead to enormous uncertainties later on. This is known as the "butterfly effect," referring to the notion that the fluttering of a butterfly's wings in Brazil could make it rain in New York a month afterwards. A change in the state of a single electron in a far-off galaxy could make a difference to events on this earth, given sufficient time. Suppose, then, that God did act by making minuscule alterations in such systems. As in the case of quantum intervention, we would not be able to prove whether he had acted or not.

Polkinghorne explicitly denies that theology, at a time when science is so dominant in our culture, needs to resort to the kind of extreme monotheism espoused by Newton. Put crudely, such a picture is not exciting enough, not shocking enough. Modern physicists are used to seeing commonsense notions violently upended by discoveries about the material world that seem almost crazy at first sight. Theories of the spiritual world need to be at least as upsetting and arresting. "I do not find," he wrote in *The Faith of a Physicist*, "that a Trinitarian and incarnational theology needs to be abandoned in favor of a toned-down theology of a Cosmic Mind and an Inspired Teacher, alleged to be more accessible to the modern mind." In fact, the modern mind is so used to being bowled over by intuition-shattering scientific theories whose main attraction is their *in*accessibility to pedestrian ways of thinking that anything failing to turn settled ideas on their head is apt to be suspect as too placid, too ordinary, to be true. A scientist, Polkinghorne stresses, expects a fundamental theory to be "tough, surprising and exciting." If it is none of those things, its value is in doubt.

The more removed God is from the world as "cosmic mind," the milder, the less extraordinary is his biography, his function as explanation of the deep mystery in which we find ourselves. But curiously, by giving him access via physical uncertainties, he is also to a large extent domesticated. He moves so discreetly, so subtly behind the scenes, we do not notice he is doing anything at all. There is a certain Jeevesian quality to this sort of divinity. And in order to be discreet, to be unobtrusive, he submits to some of the constraints we humans cannot escape. One such constraint is the one-way direction of time, from past to future, in Polkinghorne's theology.

The Bible takes God out of time completely. He is simply "I am," forever and always. Anything else, it seems, would be to create an idol. St. Augustine wrote that God created time as well as everything else, so it is a silly question to ask what he was doing prior to the Creation. As late as the nineteenth century it was said that God is not even "everlasting," since that would imply he is still in time, even if the time is of infinite duration.

Newton broke with the traditional view that time is inseparable from motion. Things "take" time to travel from one place to another. A clock measures time by the movement of its hands. But Newton proposed another sort of time, absolute time, which exists whether or not anything is moving. It is a mere "container" of objects and events. The English theologian Grace Jantzen thinks Newton must have spotted a dilemma here: how can we square this theory of time with the doctrine of the Creation, of a Beginning? A way out was to alter drastically the accepted view of the relation between God and the world. The God of

Dominion, above it all and timeless, was modified into a God whose connection with the world, while still being one of total omnipotence, is uniquely intimate and involved. He is not outside time. He *is* time. He constitutes duration as well as space, and he permeates both. He is "omnitemporal." This is where Newton got into a squabble with Leibniz, whose deity never needed to interfere in his creation since the plan of providence was all worked out and complete before the world began.

God *must* be eternal. That is a nonnegotiable, core doctrine of the Church. But once we accept that God is love as well as power, the situation is not so simple. The whole point of love is that it relates to something or someone, and that something, his creation, is always in process, never the same from one moment to the next. A standard answer is that God relates to the whole history of his creation, from Alpha to Omega, all at once. That respects the idea that God is perfect: to get entangled with time and change is a stain on perfection. But modern physics shows that our world, the universe, and everything is less "perfect," less static and set in its ways than classical theories ever hinted at.

One of the great discoveries of the twentieth century was that the universe has a history, some 15 billion years of it. But when Aquinas speaks of God already knowing the future, it is as if the world is simply space, spread out all at once, like a book whose last chapter has been written. It "is," but it does not "become." Einstein, a staunch determinist, late in his career stated that the difference between past and future is an illusion. Today, we know that time is not in the least like space, and the world is "baggy." It contains a random element, making it far less simple than Einstein's *ex cathedra* pronouncement suggests. The future is *created*; it is not just there waiting to be experienced. Here Polkinghorne drops one of his biggest theological percussion bombs: If even the omnipotent God cannot act to alter the past, it seems likely the "omniscient" God cannot know with certainty what will happen in the future, because we humans, about whose ability to know the divine Calvin made such disparaging remarks, have not "made" the future yet. It is not there to be known.

CHAPTER TWENTY-TWO

The Tiger and the Lamb

One feels inclined to say that the intention that man
should be happy is not included in the plan of creation.
—SIGMUND FREUD

A GOD WHO IS IN THE DARK as to what lies ahead matches well
the God of the Old Testament, whose biography dwells on moments of
puzzled disappointment and displeasure at the unruliness and shabby
conduct of his creatures. His omnipotence is inconsistent with the way
his expectations are dashed. As someone who thinks religion ought to
have the "toughness" of science, John Polkinghorne prefers the volatile
God of the early chapters of the Bible to the more staid and stable one
presented by classical theism. If we take time seriously—very seriously—
his partiality makes some sense. It makes even more sense if we believe,
as he does, that we are "partners" with God in inventing the future, a
concept which was rank heresy to Calvin and the Reformers.

Important limits are placed on this kind of God. He may well be able
to make highly informed conjectures about the future, and make plans
which take care of any eventuality. But his actual knowledge is open to a
future that is partly made by, not predestined for, humans. The similar-
ity to Yahweh is striking: he can suspend retribution when sinners
repent or inflict it summarily when they don't. His workings in history
are sometimes improvised. Polkinghorne reads 2 Kings 20:1–6, where

Hezekiah is first told to put his house in order to get ready to die, then given fifteen more years of life, as an example of the God of Israel having a change of mind. Yahweh is a God who responds to the prayers of his people. When Abraham pleads with God not to destroy every soul in Sodom and Gomorrah, this is not an empty charade making it seem as if God has not yet made up his mind. Prayer itself is a collaboration between the divine will and our part in realizing it. This God is, by his own wish, not in full control of the world.

Such loss of divine control, this decision to let the world be itself, is "precarious." It is a middle way in which God interacts with his creation without overruling it. As in the past, there is the risk that a God straddling the divide between the worldly and the otherworldly may be pushed too far in one direction or the other. In the excesses of "Enthusiasm," he was too personal. In the brief regime of Deism, he was too remote. Almost as if he needs to preserve God's dignity—as generations of theologians before him, not to mention the astrophysicist Lemaître, had insisted on guarding his privacy—Polkinghorne is adamant that whatever form divine interaction with history may take, it remains absolutely veiled from us humans. It will be contained within the "cloudy unpredictabilities" of the new physics, discernible by faith but impossible to prove by scientific experiment. It will, he says, "more readily have the character of benign coincidence than of a naked act of power. It will be part of the complex nexus of occurrence from which it cannot be disentangled in some simplistic way that seeks to assert that God did this but nature did that. All forms of agency intertwine."

Newton proposed the existence of a universal medium in which God's interactions with the world could take place in an orderly way. He called it "the aether." Polkinghorne is suggesting a new kind of medium, the medium of information. The study of complex systems shows they have a double nature: they are part energy, but they are also part pattern. They display a sort of orderly disorder. How such a system will evolve is not totally haphazard, being limited to a number of possible patterns of motion scientists call a "strange attractor." These patterns do not differ in the amount of energy they contain, but they do differ in the details of how they evolve. "Bottom-up" description needs to take into account the energy exchanged between the parts of the system, while a "top-down" one includes an agency of some kind that forms patterns out of possibilities left open at the lower level. There is an analogy here between mind and body in the human system and information and energy in the physical one. Polkinghorne sees "just a glimmer (no more) of the integration of the material with something that begins to look a little like the mental." He proposes that God interacts with the world through "information input" into open physical process, a selection of options. Unlike his creatures, he expends no energy. He is like "the director of the great cosmic improvisatory play, rather than an invisible actor on the stage of the universe," an "improviser of unsurpassed ingenuity."

Polkinghorne is one of a very few believing physicists who actually attempts to fit the doctrine of the Trinity into his scientific theorizing. This project is not appealing to rigorously secular scientists, in part because it assumes "the Truth" is guaranteed by supernatural agencies at a juncture in history when final explanations as to the operations of nature are elusive. But Polkinghorne has flatly stated that "Whether acknowledged or not, it is the Holy Spirit, the Spirit of Truth, who is at work in the truth-seeking community of scientists. That community's repeated experiences of wonder at the disclosed order of the universe are, in fact, tacit acts of worship of its Creator."

Recently, there has been some interesting speculation that one day it may be possible to understand all of physics in terms of information. The starting point is a remark by one of quantum theory's great pioneers, Niels Bohr, that, especially in the world of the very small, we do not describe that world itself but only what we are able to say about it. We need information as a mediator between ourselves and the subatomic realm, which cannot be envisaged in commonsense terms. Nature does not consist of matter and energy alone, as had been thought: it is matter, energy, and information. It is a mistake, the physicist Hans Christian von Bayer says, to imagine we can come to grips with the objective material world "without acknowledging, or even trying to understand, the mediating role of information." A scientist does not "see" an atom. He collects information about it and encodes it in a mathematical device called a wave function, which makes predictions about information that may be collected in future experiments.

Information in the scientific sense is measured by bits, a "yes" or "no" answer to a question. To measure the polarization of a photon, a scientist uses a counter flashing numbers that encode whether the photon did or did not arrive at a certain detector when passing through slits set at certain angles and mirrors installed at certain places. We get a "yes" or a "no." From these bits of information we can derive a statement such as "the photon is vertical 35 percent of the time." John Wheeler, the physicist who coined the term "black hole," and was a colleague and confidant of Einstein and Bohr, is one of the strongest proponents of the information thesis. He states that every particle, every field of force, even the space-time continuum itself, obtains its meaning, its very existence, even if sometimes obliquely, from an apparatus that gives a "yes" or a "no" answer to questions. "Every item in the physical world," Wheeler says, "has at bottom—at a very deep bottom in most instances—an immaterial source and explanation." What we call reality, he goes on, "arises in the last analysis from the posing of yes-no questions and the registering of equipment evoked responses. In short, that all things physical are information-theoretic in origin." His bumper-sticker slogan is IT FROM BIT.

Harold Bloom, in his role as critic of religion, has mercilessly lampooned the notion of providence as a divine input of information into an open system. The modern form of Gnosticism—a doctrine of secret knowledge of the divine which he calls the hidden religion of the United States—is "a kind of information theory." Bloom says that as Americans, "we are obsessed with information and we regard religion as the most vital aspect of information. Information becomes the emblem of salvation. In Gnosticism the false Creation-Fall concerned matter and energy, but the Pleroma, or Fullness, the original Abyss, is all information."

By aligning God with information and stressing its contrast with matter-energy, and by suggesting an analogy with mind and body, we are treading on hazardous turf that has beckoned seductively to theology more than once in the past. The theory wants God to be everywhere; but at the same time it requires him to be somewhere specific, which is ominously evocative of Newton's error, celebrating God as transcendent sovereign *and* as watchmender always on duty. As Polkinghorne says, the God of information must not be an "entity among other entities, concerning whom one can say, 'Lo here,' or 'Lo there,' thus locating him within some circumscribed religious domain. Rather, he is the source of all that is, the One omnipresent to every human experience. There is no possibility of identifying his presence by contrasting it with his absence." But that is leaning rather perilously in the direction of Deism, with its short shelf life. If we are to have a personal God, he must be in some sense local, not simply cosmic, not "the mind of the universe." Polkinghorne acknowledges this: "A personal God, such as that of which Christianity speaks, must be capable of specific response to specific circumstances." Otherwise it would be misleading even to talk about a personal relationship. Hence the God of information input.

Are we drifting into a "worst of both worlds" situation, inventing a God who has been domesticated but not making him fully personal? The American theologian Michael Buckley warned of the corrosive effect of basing a theory of religion on an intellectual premise, no matter how ingenious and impressive. What tends to ensue is not a heretical religion but an acceptance of the premise coupled with a dismissal of its religious implications. Over the centuries, says Buckley, physics, mathematics, medicine, every science, asserted their independence from theology, "and the theologians who had deposited all their coin with them found themselves bankrupt." It was on this account that Newton became "one of history's great losers." Modern atheism is a byproduct of religion's well-intentioned attempt to affirm its intellectual credentials by partnering with the most respectable intellectual enterprise going, the new science. "In an effort to secure its basis, religion unknowingly fathered its own estrangement." It was a failure of nerve. Religion lacked the stomach to stand by a core principle of its teaching, that religious knowledge is unique, different, untranslatable into the language of any secular form of knowing. If religion has no intrinsic justification, it cannot be justified from the outside.

An information theory God is a God of intellect rather than of will: will is associated with the energy he does not need to use. At once, he is scaled down to more human size, to a more rational style. He is less foreign, more familiar. We can talk about him in terms not radically other than those we use to characterize a colleague or friend. Polkinghorne occasionally lapses into this kind of talk. He calls God an improviser, gracious, a lover, a theatrical director, a sharer, vulnerable, reliable, inventive, creative, an economist, a being who lives with "precariousness." There is an echo here of the seventeenth-century "transparency" that led brilliant minds to "get God wrong."

Paul Tillich thought the doctrine of God as will makes the world seem incalculable, treacherous, unsafe. The laws of nature are orderly and rational, but behind them stands the possibly irrational force of will, which could smash the order and throw it into confusion in the blink of an eye. The stability of the universe, as well as the salvation of our souls, are not necessarily the case. It must be said that the God of the Old Testament, impossible to ignore in any honest discussion of the relation between the human and the divine, certainly seems to be a deity in which the will is dominant.

The Old Testament scholar David Penchansky talks about the "sinister" image of God that emerges in parts of the Hebrew Bible. This, he believes, is not just a primitive residue of an earlier, not entirely monotheistic religion, but a genuine expression of an Israelite theology. The reasonable, civilized, and discreetly tactful divinity of today's science does not begin to do justice to what actually happens in the world of the deeply religious authors of the dark biblical masterpieces. Where is God the enemy; where is the driving force of history who breaks out of the neat categories of our human ethics and ideas of right and wrong with unimaginable drama and violence?

The Book of Job, Penchansky thinks, is a profoundly complex inquiry into the riddle of such a divinity who cannot be made to fit the normal requirements of decent conduct by trying to harmonize his disparate traits. Job himself, reeling under the impact of crushing misfortune, a social and financial outcast, knows he must protest; but it is hopeless to do so in the style of a civilized family man who prides himself on his respectable place in society. His remonstration is, perforce, not an ordered indictment of the Almighty "but rather an unrestrained shriek, a howl of pain." The incomprehensible is opposed by the inarticulate.

One of Job's problems is that he has no mediator to negotiate or intercede with a remote, sometimes obtuse God who does not, for all his bluster about creating a fine and splendid universe, seem to be in full control. Seeking a go-between and not locating one, Job finds God all the more intractable and impossible to understand. He is trying to make human sense of a divinity that is incompatible with the grim realities of human life. By ignoring the horrors of a world in which the innocent and virtuous are victims of the worst bad fortune can throw at them, we can maintain a portrait of an omnipotent, omniscient, and benevolent God, the God of Newton. But Job's plight makes that difficult. He cannot alter his sorry situation. The only recourse for a serious theology is to attempt a new concept of God, one that faces up to certain squalid and uncomfortable facts. There seem to be not one but two aspects of God in the Book of Job, as well as a Satan who at times is more powerful, more resolute and single-minded than the king of heaven. There may be, Penchansky theorizes, a residue of Canaanite polytheism and Persian dualism here, systems which permit other supernatural beings to take the blame for the suffering humans must endure. But polytheism never worked well in the long run. In Job, there is a glaring absence of any reading of God able to make sense of what is happening to his afflicted servant.

That is why Job's story can be relevant to the age of the holocaust, the Twin Towers, the 2004 tsunami. But it is germane only if we readers recognize in it "some divine responsibility for human evil, human stupidity and humanity's self-destructive tendencies. There remain no more effective theological subterfuges to maintain a defense against the reality of worldwide human pain and death, if indeed there ever were any such." A reinvention of classical theism is needed, which is outside the competence of modern physics, however ingenious and well meaning, however consoling it may be to have a God whose mind moves in the same grooves as our admired maestros of chaos and quantum theory. Science has not been laggard in reinventing the Creator. Now, in the opinion of many, theology has to undertake a similar enterprise, though in its own unique and inimitable way.

Critics of the Book of Job have attempted to force the story into a consistent unity, reflecting the actions of a single, omnipotent God in a coherent and unified universe. All such efforts have met with varying degrees of failure. Job does accept his misfortune early on, and in the end God heals him. But the text subverts any idea that piety is rewarded as a matter of course. A question mark is placed over accepted notions of what sort of God we believe in. Rather than plead his virtuousness, Job instead asserts his integrity. And now, as Penchansky makes clear, we see that integrity is something different from piety, from reverence for an incomprehensible deity. Integrity is more human than divine. It is, in fact, at odds with God, if we read the text of Job correctly. It is a sort of painful recognition of the brute fact of personal unhappiness and the fragility of what were once imagined as rock-steady principles baked into the cake of the universe.

The new concept of piety does not mean adherence to the rules of religious authority. There are too many awkward incongruities for that to apply any more. Integrity, something hardwon in the face of unreasonable affliction, is "loyalty to one's own experience," however awful and ugly that experience may be.

Where in this loyalty to oneself does loyalty to God, the God who is the hero-villain of the Book of Job, come in? Penchansky is fairly blunt on this point. "One might even wonder," he says, "whether such a God, who bullies, blusters and overpowers a human questioner is worthy of the respect offered him, whether gaining the approval of such a God should be important to a person of integrity." The perplexity of God's ambiguity, conveyed so powerfully in the text, perhaps hints at the breakup of the cult of ancient Yahwism, imported from Babylon, with its simple theory of a transcendent God in full control. No longer was it taken for granted that God underwrites the social and cosmic order. Job's casting about for a mediator, even though he does not find one, seems to weaken the strict monotheism of Israel.

A lesson of Job, Penchansky surmises, is that a new image of God is needed, one based on the fact of dissonance in the world. The harmonies of Newton's celestial mechanics for generations helped to mask the grimmer implications of discord. Dissonance may be what the Book of Job is about. A reinvention of God that ignores this given of our human situation would be mere theological wheel-spinning, harking back to a model that no longer fits reality. Such a reconstruction would accept the absence of any reliable "relationship of obligation" between God and his people, but admit a faithful trust *in spite of* an acknowledgment of Job-like experiences, which could suggest "divine enmity, God's moral failure." That would mean a less triumphal, less authoritarian deity, and, as a corollary, an affirmation of the need to take control of human integrity, which is not a gracious gift of God or the universe but something requiring great individual fortitude. One way or another there will be suffering, and that must be part of any theology worth the name.

On Good Friday, 1966, *Time* magazine published an article entitled "Is God Dead?" His suspected assassin was modern science, according to *Time*, in comparison with which everything else seems uninteresting and unreal. The magazine quoted a young Catholic philosopher, Michael Novak, as saying: "If, occasionally, I raise my heart in prayer, it is to no God I can see, hear or feel." In September 2005, *Newsweek* announced a massive, exuberant upsurge of spiritual questing in America, a hunger for a direct experience of the divine. Michael Novak, now seventy-two, was asked if he would like to correct the record on his youthful sense of an absent God. Novak replied that God is as far away today as he has ever been. Religion, he said, can seem to be a jubilant celebration of the divine, as it evidently does to millions of "spiritual" Americans. That is fine when life is going swimmingly and God seems to be showering one with blessings. Job was serenely comfortable for years. But the real test of faith, Novak stressed, is in adversity and despair, when God does not show up in every blade of grass or storefront church. "That's when the true nature of belief comes out," he said. "Joy is appropriate for the beginnings of your faith. But sooner or later somebody will get cancer, or your best friend will betray you. That's when you will be tested."

In the Old Testament, the Israelites did move toward a different concept of God, but it was a retreat to the traditional picture of a big, dominant being of majesty and judgment. The misfortunes of the people were the inevitable result of disobedience. Instead of acknowledging the divine mystery, God was defined in more specific ways than hitherto. Far from growing diffident or tentative, his power expanded, became world-transcending. Life is unfair, it seems, and the virtuous seem to be cruelly mistreated. But wait. Today's injustice will be corrected. Not now, not tomorrow, but in the indefinite future. In that way one could make sense of the great enigma of a good God ruling a wicked world. As a result, "Israelite pain was driven inwards."

There are always two options. If a picture of God is dissonant with the situation of our actual experience, we can either change the picture or reinterpret the experience. The Israelites chose the latter alternative, but in doing so were false to themselves. They had not learned the lesson of integrity that Job taught. They could not accept that there is *unexplained* menace in a world that is not harmonious in all respects. Penchansky turns to the famous poem of William Blake as a commentary on this hard but unavoidable reality:

Tyger, Tyger, burning bright,
In the forest of the night.
What immortal hand or eye,
Could frame thy fearful symmetry?

In what distant deeps or skies
Burnt the fire of thine eyes?
On what wings dare he aspire?
What the hand dare seize the fire?

And what shoulder, & what art,
Could twist the sinews of thy heart?
And when thy heart began to beat,
What dread hand? & what dread feet?

What the hammer? What the chain?
In what furnace was thy brain?
What the anvil? What dread grasp,
Dare its deadly terrors clasp?

When the stars threw down their spears
And watered heaven with their tears:
Did he smile his work to see?
Did he who made the Lamb make thee?

There is no one image of God that works in all circumstances. The domesticated lamb and the rapacious tiger are part of the same world, which in monotheism is the creation of one and the same Maker. The anthropic principle is hopelessly inadequate to come to grips with such a deep mystery, as are all theories of divine tinkering in the crevices of physical uncertainty. The Old Testament writers would have laughed at such inventions. As an *envoi* to the nightmare events of the past hundred years, Penchansky suggests we need to revisit the notion of a dangerous God, "perhaps even an evil God." There is ample precedent in the Hebrew scriptures. The stories told there of irrational episodes in the career of the Old Testament God "might provide a structure for processing the evil in the world." That would entail an amicable separation between the two realms of theology and science.

In the early days of Darwinism, the nineteenth-century scholar Herbert Spencer wrote that religions tend to harbor a secret fear that everything may some day be explained, which suggests they are hiding a residual doubt as to whether God as an Incomprehensible Cause is really as incomprehensible as they supposed. What they must face up to, Spencer said, is that it is only in the assertion of a reality utterly inscrutable that religion can be reconciled with science. "A permanent peace between science and religion," he said, "will be reached when science becomes fully convinced that its explanations are proximate and relative, while religion becomes fully convinced that the mystery it contemplates is ultimate and absolute."

The first condition has arguably been met. The second may be a long time in coming.

Notes

PROLOGUE

xiii   "Theology necessarily is a system": Harold Bloom, *Jesus and Yahweh: The Names Divine* (New York: Riverhead Books, 2005), p. 98.

xiii   "The style of thy works": John Donne, *Devotions Upon Emergent Occasions and Death's Duel* (New York: Vintage Spiritual Classics, 1999), p. 119.

xiv   a "Copernicus in poetry": John Carey, ed., *John Donne: The Major Works* (Oxford and New York: Oxford University Press, 1990), p. xix.

xiv   "a pack of fools": William R. Shea and Mariano Artigas, *Galileo in Rome* (Oxford and New York: Oxford University Press, 2003), p. 187.

xiv   "He that seeks proofe": Quoted in John Moss, ed., *One Equall Light* (Grand Rapids, MI, and Cambridge: William B. Eerdmans, 2003), p. 145.

xiv   to "batter heaven": Jeffrey Johnson, *The Theology of John Donne* (Cambridge: D. S. Brewer, 1999), p. 47.

xiv   "Prayer . . . hath the nature of Impudency": Ibid., p. 47.

xv   "so steepy a place to clamber up": Ibid., p. 6.

xvi   "born a priest of nature": *Works of the Honourable Robert Boyle* (London, 1772) quoted in David W. Noble, *The Religion of Technology: The Divinity of Man and the Spirit of Invention* (New York and London: Penguin Books, 1997), p. 64.

xvii   "I'm not very optimistic": John Rodwell in *Faith in Science*, ed. W. Mark Richardson and Gordy Slack (London and New York: Routledge, 2001), p. 47.

xvii   "This is especially true": Brian Cantwell Smith in ibid., p. 68 (italics in the original).

xviii   He has to break us down: Søren Kierkegaard, *Journals*, in Alexander Dru, ed., *The Soul of Kierkegaard* (Mineola, NY: Dover Publications, 2003), p. 245.

xix opaque, even "sinister" God: David Penchansky, *What Rough Beast?* (Louisville, KY: Westminster John Knox Press, 1999), p. 94.

CHAPTER ONE

3 "And so it will be": Fyodor Dostoevsky, *The Brothers Karamazov*, trans. Richard Pevear and Larissa Volokhonsky (San Francisco: North Point Press, 1990), p. 254.

3 "This idea that the human mind": Paul Tillich, *A History of Christian Thought* (New York: Simon & Schuster, 1968), p. 264.

3 the "sweet deceits": Steven Ozment, *Protestants: The Birth of a Revolution* (New York and London: Image Books, Doubleday, 1992), p. 217.

4 "an impious lie": John Calvin, *Institutes of the Christian Religion,* trans. Henry Beveridge, Vol. 1 (Grand Rapids, MI: William B. Eerdmans, 1973), p. 91.

4 a "foolish and inconsiderate longing": Ibid., p. 100.

4 "Daily experience shows": Ibid., p. 97.

4 "Coming into the Church of Our Lady": Andrew Pettegree, *Europe in the Sixteenth Century* (Oxford and Malden, MA: Blackwell, 2002), p. 170.

5 The authorities were somewhat abashed: Carlos Eire, *War Against the Idols: The Reformation of Worship from Erasmus to Calvin* (Cambridge and New York: Cambridge University Press, 1986), pp. 115–19.

6 When a gang of incendiarists broke into a church: Christopher Haigh, *English Reformations* (Oxford and New York: Clarendon Press, 1993), p. 69.

6 "images and adorning of them": Eamon Duffy, *The Voices of Morebath: Reformation and Rebellion in an English Village* (New Haven and London: Yale University Press, 2001), p. 93.

7 "peremptory abhorrers of Pictures": Johnson, *The Theology of John Donne*, p. 65.

7 played lewd tunes on the organ: Antonia Fraser, *Cromwell* (New York: Grove Press, 1973), p. 103.

8 "the Pope's Triple Crowne": John Morrill, "Sir William Brereton and England's Wars of Religion," in Peter Gaunt, ed., *The English Civil War* (Oxford and Malden, MA: Blackwell, 2000), p. 192.

8 "gaudy glistenings": Christopher Hill, *Milton and the English Revolution* (New York: Viking, 1978), p. 174.

8 to "leap and fly": Francis Bacon, *Novum Organon*, trans. and ed. Peter Urbach and John Gibson (Chicago and La Salle, IL: Open Court, 1994), p. 48.

9 called "idols": Ibid., pp. 53–74.

9 "need to be smashed": John Channing Briggs, "Bacon's Science and Religion," in Marrku Peltonen, ed., *The Cambridge Companion to Bacon* (Cambridge and New York: Cambridge University Press, 1996), pp. 177–78.

9   "For let men please themselves": Bacon, *Novum Organon,* p. 23.

9   a mind that is not entirely naked: See Jeremy Campbell, *The Improbable Machine* (New York and London: Simon & Schuster, 1991), p. 125.

10   "Men emancipated themselves": Christopher Hill, *The World Turned Upside Down* (London and New York: Penguin Books, 1991), p. 124.

11   it amounted to a kind of polytheism: Eire, *War Against the Idols,* p. 12.

11   "developed a nebulous concept of divinity": Ibid.

12   "In practical terms": Ibid., p. 316.

12   "the new papists": Ibid., p. 313.

12   "Those who broke the law": Ibid., p. 317.

13   a "theology of mediation": John MacQuarrie, *The Mediators* (London: SCM Press, 1995), p. 144.

CHAPTER TWO

14   "a taste for opposites": Tertullian of Carthage, *Against Marcion,* 1.16, quoted in Eric Osborn, *Tertullian, First Theologian of the West* (Cambridge and New York: Cambridge University Press, 1997), p. 94.

14   "It is narrated of the biologist Haekel": James H. Breasted, *The Dawn of Conscience* (New York and London: Charles Scribners Sons, 1954), p. 387.

15   "may be thought of us to be": Thomas Digges, *A Perfect Description of the Celestiall Orbes* (1576), printed in Dennis Richard Danielson, ed., *The Book of the Cosmos: Imagining the Universe from Heraclitus to Hawking* (Cambridge, MA: Perseus Publishing, 2000), p. 137.

15   God's "infinite excellence": Ibid., pp. 141–42.

15   "We have not now any Drake or Columbus": Ibid., p. 190.

15   the vastness and silences of interstellar space: Blaise Pascal, *Pensées and Other Writings,* trans. Honor Levi (Oxford and New York: Oxford University Press, 1995), p. 73

16   "When the sky was only": Bernard le Bovier de Fontenelle, *Conversations on the Plurality of Worlds,* trans. H. A. Hargreaves (Berkeley and London: University of California Press, 1990), p. 63.

16   "it may be allow'd that God": Isaac Newton, *Opticks* (1704) (New York: Dover Publications, 1979), pp. 403–4.

16   "subtle men, with some appearance of probability": Carey, ed., *John Donne: The Major Works,* p. 391

16   Nature is "inexorable and immutable": Stillman Drake, trans., *Discoveries and Opinions of Galileo* (New York and London: Anchor Books, Doubleday, 1957), p. 182.

17   Job who seems to have rendered God speechless: Jack Miles, *God, A Biography* (New York: Vintage Books, 1996), p. 11.

17   "Divine neglect": James Kugel, *The God of Old* (New York and London: Free Press, 2003), p. 61.

18   "The same God who buttonholes the patriarchs": Ibid., p. 61.

18   "That is the God": Ibid., p. 63.

19   "Messy monotheism": Jack Nelson-Pallmeyer, *Jesus Against Christianity* (Harrisburg, PA: Trinity Press International, 2001), p. 77.

20   "The designation of God as Almighty": Ibid., p. 296.

23   One seraphim cries out to another: Isaiah 6:3.

23   "Contrary to popular misconception": Gerhard von Rad, *The Message of the Prophets* (New York and London: Harper & Row, 1967), p. 38.

23   The seraphs . . . were really announcing: Karen Armstrong, *A History of God* (New York: Alfred A. Knopf, 1993), p. 41.

24   "bear the brunt of the divine impact": Ibid., p. 56.

24   "The prophets in an important sense": Ibid., p. 39.

24   "All religion must begin": Ibid., p. 48.

25   "The Lord may exalt the righteous": Miles, *God, A Biography*, p. 339.

26   "If God is capable of testing mankind": Ibid., pp. 332–33.

27   a "stiffnecked people": Exodus 32:9.

27   "Remember, and forget not": Deuteronomy 9:7.

27   "It is not just that God is good": Richard Elliott Friedman, *The Disappearance of God: A Divine Mystery* (Boston and New York: Little, Brown, 1995), p. 109.

28   as God's "architect": See Proverbs 8:22–31.

28   An influential scholar, Wilhelm Bousset: See Larry W. Hurtado, *One God, One Lord* (London and New York: T. & T. Clark, 1998), p. 23.

28   by speaking the names of angels: Ibid., p. 35.

28   A very ancient Jewish tradition: See Christopher Rowland, *Christian Origins* (London: SPCK Press, 1985), p. 31.

28   Michael was singled out: Daniel 12:11.

28   a grand vizier was given authority: Hurtado, *One God, One Lord*, p. 83.

CHAPTER THREE

29   "Considering therefore the signification": Thomas Hobbes, *Leviathan* (1651) (London and New York: Penguin Books, 1985), pp. 439–40.

29   "Are we only a parody of the angels?": Harold Bloom, *Omens of Millennium* (New York: Riverhead Books, 1997), p. 14.

29   "The spiritual life": Ibid., p. 17.

30   "There is a primitive knowledge in man": Joseph Cardinal Ratzinger, *God and the World: A Conversation with Peter Seewald* (San Francisco: Ignatius Press, 2002), p. 122.

30   developed "anti-angel aggression": Flannery O'Connor, *The Habit of Being*, quoted in Harriet Chessman, *Literary Angels* (New York: Ballantine Books, 1994), pp. 104–5.

33   James Kugel sees angels taking up: Kugel, *The God of Old*, p. 193.

33   "Toward the end of the biblical period": Ibid., p. 194.

34 a watershed chapter in the history of religion: Tillich, *A History of Christian Thought*, p. 10.

36 "For a brief and exciting period": Stephen M. Fallon, *Milton Among the Philosophers* (Ithaca, NY, and London: Cornell University Press, 1991), pp. 138, 167.

36 "divested of their sublimity": Bloom, *Omens of Millennium*, p. 43.

37 "the eternal misunderstanding between God and man": Max Brod, *Franz Kafka: A Biography* (New York: Da Capo Press, 1994), p. 174.

37 "Nor may we invoke Angels": Quoted in Frank E. Manuel, *The Religion of Isaac Newton* (Oxford: Clarendon Press, 1974), p. 60.

CHAPTER FOUR

39 "As a lifelong critic": Bloom, *Jesus and Yahweh*, p. 99.

40 the "controlled despair": Tillich, *A History of Christian Thought*, pp. 12–13.

41 "faint insinuation of a Trinitarian dimension": Quoted in William La Due, *The Trinity Guide to the Trinity* (Harrisburg, PA: Trinity Press International, 2003), p. 11.

42 found no evidence that he was called God: Quoted in ibid., p. 21.

42 the hero cults of the ancient world: Tillich, *A History of Christian Thought*, p. 71.

45 took on an intimidatingly technical and abstract tone: Catherine Mowry LaCugna, *God for Us: The Trinity and Christian Life* (San Francisco: Harper & Row, 1991), p. 2.

45 "contempt which has for many years been cast": Quoted in Roland N. Stromberg, *Religious Liberalism in Eighteenth Century England* (London and Oxford: Oxford University Press, 1954), p. 48.

45 to keep it simple: See Margaret Jacob, *The Newtonians and the English Revolution* (Hassocks, Sussex: Harvester Press, 1976), p. 46.

45 obstinately held to the position: See Karl Barth, *Dogmatics in Outline*, trans. G. T. Thomson (London: SCM Press, 1960), p. 40.

46 thinks the Trinity doctrine could be: See Jürgen Moltmann, *God in Creation: The Gifford Lectures 1984–1985* (Minneapolis: Fortress Press, 1993), p. 86.

46 the awfulness of the godhead: Helen Gardner, *Religion and Literature* (London: Faber & Faber, 1971), pp. 176–77.

47 Frederick the Wise ... assembled a fabulous collection: Steven E. Ozment, *The Reformation in the Cities* (New Haven and London: Yale University Press, 1975), p. 139.

47 "Don't touch the relics!": Geoffrey Parker, *Success Is Never Final* (New York: Basic Books, 2002), pp. 31–32.

48 "Omnipotence cannot be exalted": Quoted in Gardner, *Religion and Literature*, p. 123.

CHAPTER FIVE

50   "The whole Frame of the world": John Donne, Sermons VIII, 9.176–77, quoted in Moss, ed., *One Equall Light*, p. 144.

50   an Oriental tale: Gustav Davidson, *A Dictionary of Angels* (New York: Free Press, 1967), pp. 193–94.

51   "higher than heaven, deeper than the abyss": Quoted in Jarislav Pelikan, *The Christian Tradition*. Vol. 3: *The Growth of Medieval Theology* (Chicago and London: Chicago University Press, 1978), p. 70.

51   "render thy spouse, the most blessed Virgin": Ibid., p. 42.

51   Myths and outright frauds were used: Bart D. Ehrman, *The Lost Christianities* (Oxford and New York: Oxford University Press, 2003), p. 207.

52   The midwife rushes out to bring: Ibid., p. 209.

53   "a millstone about the neck": J. S. Whale, *The Protestant Tradition* (Cambridge: Cambridge University Press, 1960), p. 122.

53   He did not like this modish bypassing: H. George Anderson, J. Francis Stafford, and Joseph A. Burgess, *The One Mediator, the Saints and Mary* (Minneapolis: Augsburg, 1992), p. 25.

53   A Catholic reply . . . couched in a harsh idiom: Ibid., p. 28.

54   He blamed women . . . for the mess the world was in: Charlene Spretnak, *Missing Mary* (New York and Basingstoke, Surrey: Palgrave Macmillan, 2004), p. 151.

54   fears of imitating the old goddess cults: Ibid., p. 152.

55   "All subsequent elevations": Ibid., p. 153.

55   brings Mary down to earth: Mary Hines, *Whatever Happened to Mary?* (Notre Dame, IN: Ave Maria Press, 2001), pp. 162–63.

56   watched the closing ceremony . . . "through their tears": Ibid., p. 166.

57   "a childlike gaiety and optimism": A. G. Dickens, *The English Reformation* (New York: Schocken Books, 1964), p. 6.

57   purgatory, as a "third zone": Jacques Le Goff, *The Birth of Purgatory*, trans. Arthur Goldhamme (Chicago: University of Chicago Press, 1981), p. 41.

58   bringing "air or a sweet fragrance": Ibid., p. 137.

58   "more horrible than the pain itself": Dickens, *The English Reformation*, p. 5.

58   as an example of a religious practice: Ibid., p. 6.

59   "standing altarwise": Haigh, *English Reformations*, p. 24.

59   "brought a strange new burden to bear": Steven Ozment, *The Age of Reform, 1250–1550* (New Haven and London: Yale University Press, 1981), p. 437.

CHAPTER SIX

60   "Any idea of God": Stephen Mitchell, *The Book of Job* (New York: Harper Perennial, 1987), p. xiv.

60   "It is remarkable": Tillich, *A History of Christian Thought*, p. 270.

60   "foundered on man's indomitable credulity": Ozment, *The Age of Reform, 1250–1550*, p. 438.

60   "To the fear of hell": Carl Bridenbaugh, *Vexed and Troubled Englishmen, 1590–1642* (Oxford and New York: Oxford University Press, 1967), p. 276.

61   that the new Protestant sense of God's opaqueness: Richard Marius, *Martin Luther* (Cambridge, MA, and London: Harvard University Press, 1999), p. 103.

61   "almost like a contagious disease": Tillich, *A History of Christian Thought*, p. 229.

62   "a singularly anxious man": William Bouwsma, *John Calvin: A Sixteenth Century Portrait* (Oxford and New York: Oxford University Press, 1988), p. 32.

62   Hobbes described himself and his fear as twins: See Johann P. Somerville, *Thomas Hobbes: Political Ideas in Historical Context* (New York: St. Martin's Press, 1992), p. 5.

62   "covenants entred into by fear": Hobbes, *Leviathan*, p. 107.

62   "By comparison with traditional practice": Ozment, *Protestants: The Birth of a Revolution*, p. 216.

63   "succeeded in making God more mysterious": Marius, *Martin Luther*, p. 103.

63   "For Calvin the central doctrine": Tillich, *A History of Christian Thought*, pp. 202–3.

64   "A strange thing happened": Conyers, *The Long Truce*, p. 202.

64   "power politics edging out the idea of the good society": Ibid., p. 58.

65   although the theory of divine right may seem bizarre: Richard Dunn, *The Age of Religious Wars, 1559–1715* (New York and London: W. W. Norton & Company, 1979), p. 18.

67   God "foresees, purposes, and does": Martin Luther, *Bondage of the Will*, in John Dillenberger, ed., *Martin Luther: Selected Writings* (New York: Anchor Books, 1962), p. 181.

67   "often the messenger of bad news": Calvin, *Institutes of the Christian Religion*, Vol. 1, p. 474.

67   and for making a contribution to the fear of demons: Hans Küng, *Christianity, Essence, History and Future* (New York and London: Continuum Press, 1996), pp. 296–97.

67   that the reason God does not save: Calvin, *Institutes of the Christian Religion*, Vol. 2, p. 214.

68   "so that any of their petty saints": Sermon 45 on Deuteronomy, quoted in Bouwsma, *John Calvin*, p. 164.

68   "the right perception of any matter": Franz Kafka, *The Trial*, trans. Willa and Edwin Muir (New York: Schocken Books, 1992), p. 216.

CHAPTER SEVEN

69 "Woe to him who seeks": Herman Melville, *Moby Dick or, The Whale* (New York and London: Penguin Books, 1992), p. 54.

69 "I know there is wilde love": Quoted in William C. Placher, *The Domestication of Transcendence: How Modern Thinking About God Went Wrong* (Louisville, KY: Westminster John Knox Press, 1996), p. 6.

71 "purge the world of spirits": John Hedley Brooke, "Science and the Fortunes of Natural Theology: Some Historical Perspectives," *Zygon*, vol. 24, no. 1 (March 1989), p. 8.

72 "the dominion of God was primary": Richard Westfall, *Never At Rest: A Biography of Isaac Newton* (Cambridge: Cambridge University Press, 1998), pp. 354–55.

72 "one of history's great losers": B. J. T. Dobbs, "Newton as Final Cause and First Mover," *Isis*, 85 (1994), p. 643.

73 a "dramatically different era": Placher, *The Domestication of Transcendence*, p. 1.

73 "led theology astray": Ibid., p. 3.

74 "A God with a definable location": Ibid., pp. 134–35 (italics in the original).

75 "We have received a sign": E. B. White, *Charlotte's Web* (New York: HarperCollins, 1980), p. 80.

76 "the absolute medieval commitment": Amos Funkenstein, *Theology and the Scientific Imagination* (Princeton: Princeton University Press, 1986), p. 25.

77 "has a Judaic ring to it": Richard H. Popkin, "Newton and Maimonides," in *A Straight Path: Essays in Honor of Arthur Hyman and Ruth Link-Salinger* (Washington, DC: Catholic University of America Press, 1988), p. 216.

78 "Whether we are talking about": Kenneth Seeskin, *Maimonides: A Guide for Today's Perplexed* (West Orange, NJ: Behrman House, 1991), p. 30.

78 "It is as if God is a black box": Ibid., p. 36.

78 occurs six times in four sentences: Isaac Newton, *The Principia*, trans. I. Bernard Cohen and Anne Whitman (Berkeley: University of California Press, 1999), p. 941.

79 "depends every moment on some": Quoted in Thomas C. Pfizenmaier, *The Trinitarian Theology of Dr. Samuel Clarke, 1675–1729* (Boston: Brill Academic Publishers, 1997), p. 83.

79 "We must believe that": Yahuda MS, 15.3.f 46v, quoted in ibid., p. 827.

CHAPTER EIGHT

81 "George Bernard Shaw defines": Edmund Fuller, *George Bernard Shaw: Critic of Western Morale* (New York and London: Charles Scribner's Sons, 1950), pp. 38–39.

82   required by the "richness of nature": Fantoli, Galileo, *Dialogue Concerning the Two Chief World Systems* (Berkeley: University of California Press, 1967), p. 101.

83   that mathematics was a diabolical art: Annibale Fantoli, *Galileo: For Copernicanism and for the Church* (Rome: Vatican Observatory Publications, 1996), p. 175.

84   "showed that no less praiseworthy": Giorgio de Santillana, *The Crime of Galileo* (New York: Time Inc., 1962), p. 135.

85   a "more solid doctrine": Galileo, *Dialogue Concerning the Two Chief World Systems,* p. 138.

85   "murmured in the pontiff's ear": Shea and Artigas, *Galileo in Rome,* p. 142.

86   visible to "brutes and the vulgar": Letter to Christina in Drake, *Discoveries and Opinions of Galileo,* p. 196.

86   "'Tis now evident": William Whiston, *A New Theory of the Earth* (1708), quoted in Richard Olson, *Science Deified and Science Defied* (Berkeley and London, University of California Press, 1995), p. 113.

87   "Gravity must be caused by an agent": Ibid., p. 115.

87   "Sir Isaac Newton has advanced": Historical MSS Commission, Portland Manuscripts, Norwich, 1899, v. 550, quoted in Robert Hugh Kaargon, *Atomism in England from Hariot to Newton* (Oxford: Clarendon Press, 1966), p. 136.

88   "I desired that you would not": Isaac Newton to Richard Bentley, February 25, 1692, in *Isaac Newton's Papers and Letters on Natural Philosophy* (Cambridge, MA: Harvard University Press, 1978), p. 302.

88   "There exists an infinit and omnipresent spirit": John Brooke, "The God of Isaac Newton," in John Fauvel et al., eds., *Let Newton Be!* (Oxford and New York: Oxford University Press, 1990), p. 172.

88   "more subtile and noble": Jan Golinski, "The Secret Life of an Alchemist," in ibid., p. 151.

88   his ideas about the vegetable spirit: Ibid., p. 153.

89   "bringing forth the beginning": Ibid., p. 160.

89   "besides what is owing to those active Principles": Newton, *Opticks,* pp. 399–400.

89   "what the vulgar call miracles": John Henry, "Newton, Matter and Magic," in Fauvel et al., *Let Newton Be!,* pp. 136–37.

90   "certain aetherial spirits or vapours": Ibid., p. 143.

90   "the aether is but a vehicle": Ibid., p. 143.

90   "Life and will are active principles": Quoted by Gary B. Deason, "Reformation Theology and the Mechanistic Conception of Nature," in David C. Lindberg and Ronald L. Numbers, eds., *God and Nature* (Berkeley: University of California Press, 1986), p. 184.

91   "points to the necessity of God": Ibid., p. 186.

92   the character of God underwent an alteration: Ibid., p. 187.

92   "As the key to the meaning": Ibid., pp. 185, 187.

CHAPTER NINE

93 "God being therefore hidden": Blaise Pascal, *Pensées*, trans. A. J. Krailsheimer (Harmondsworth, Middlesex, and New York: Penguin Books, 1995), pp. 63–64.

93 "In the Incarnation": Karl Barth, *Epistle to the Romans* (Oxford and New York: Oxford University Press, 1968), p. 98.

94 "is the notion of materialism and fate": H. G. Alexander, ed., *The Leibniz-Clarke Correspondence* (Manchester: Manchester University Press, 1956), p. 14.

94 "not so much by the Will": John Hedley Brooke, *Science and Religion: Some Historical Perspectives* (Cambridge and New York: Cambridge University Press, 1991), p. 161.

94 "holy wars went out with *Saint* Louis": Bertrand Russell, *History of Western Philosophy* (New York: Simon & Schuster, 1958), p. 582.

95 not to "give up washing": Ibid.

95 "Sir Isaac Newton and his followers": Alexander, *The Leibniz-Clarke Correspondence*, pp. 11–12.

96 "God has foreseen everything": Ibid., p. 18.

96 " 'Tis in the frame of the world": Samuel Clarke in ibid., p. 23.

97 the code name "magnesia": Betty Jo Teeter Dobbs, *The Janus Faces of Genius: The Role of Alchemy in Newton's Thought* (Cambridge and New York: Cambridge University Press, 1991), p. 29.

97 "Couriers from the beyond": Ibid., p. 95.

97 is his "contractor": Ibid., p. 36.

98 began in earnest with the late medieval debate: Richard Wentz, "The Domestication of the Divine," *Theology Today,* vol. 57, no. 1 (April 2000), p. 25.

99 "The whole world natural": H. McLachlan, ed., *Sir Isaac Newton's Theological Manuscripts* (Liverpool, 1950), quoted in Jacob, *The Newtonians and the English Revolution*, p. 137.

99 "a frequent and habitual": Alexander, *The Leibniz-Clarke Correspondence*, p. 195.

99 "the necessities of the marketplace": Jacob, *The Newtonians and the English Revolution*, p. 177.

99 "Thus the wise Governour": Quoted in ibid., p. 180.

100 "William Blake was not being": Jacob, *The Newtonians and the English Revolution*, p. 19.

100 "men of wide swallow": Quoted in Pfizenmaier, *The Trinitarian Theology of Dr. Samuel Clarke*, p. 44.

101 "equal to mathematical certainty": John Locke, *An Essay Concerning Human Understanding* (1690) (London: George Routledge & Sons, 1909), p. 528.

101 "to imagine that nature": Jacob, *The Newtonians and the English Revolution*, p. 18.

102 "is the clothing of God": Ibid., p. 27.

102 "is running up like parchment in the fire": Quoted in Hill, *The World Turned Upside Down*, p. 12.

103 a "third culture": Hill, *Milton and the English Revolution*, p. 69.

103 "sit still, in submission and silence": Barry Reay, *The Quakers and the English Revolution* (New York: St. Martin's Press, 1985), p. 17.

103 "out of this confusion": Hill, *The World Turned Upside Down*, p. 155.

103 "tumbling the earth upside down": Ibid., p. 158.

104 "When I wrote my treatise": Isaac Newton, Keynes MS, 130.6, quoted in Frank Manuel, *A Portrait of Isaac Newton* (New York: Da Capo, 1990), p. 125.

104 The "designer" role of God: Margaret Jacob, "Christianity and the New World View," in Lindberg and Number, *God and Nature*, p. 247.

105 "The assimilation of Newtonian science": Ibid., p. 249.

105 "in evaluating Newton": Dobbs, "Newton as Final Cause and First Mover," p. 643.

CHAPTER TEN

106 "What I have learned about God": Joan D. Chittister, "God Become Infinitely Larger," in Marcus Borg and Ross Mackenzie, eds., *God at 2000* (Harrisburg, PA: Morehouse Publishing, 2000), p. 81.

106 "There's this great image of Moses": Mitchell P. Marcus in W. Mark Richardson et al., eds., *Science and the Spiritual Quest* (New York and London: Routledge, 2002), p. 106.

107 Science "is a religion": Thomas Sprat, *The History of the Royal Society of London* (1667) (Whitefish, MT: Kossinger Publishing, 2001), p. 80.

107 "the blind applause of the ignorant": Ibid., p. 349.

107 "The infinite pretences to *Inspiration*": Ibid., p. 25.

107 "the great a-do which has been made": Ibid., pp. 25–26.

108 "subtil, refin'd and Enthusiastical": Ibid., p. 377.

108 "the satisfaction of breathing a freer air": Ibid., p. 53.

109 "commonly an unaffected sincerity": Ibid., p. 114.

109 "most considering Men": Ibid., p. 63.

110 "channels of spiritual power": Keith Ward, *Religion and Revelation* (Oxford: Clarendon Press, 1994), p. 77.

110 "mediators of the divine": Ibid., p. 72.

111 "a riotous plurality": Ibid., p. 62.

112 "Instead of thinking of God": Ibid., p. 24.

112 "posted an army of ministering spirits": Daniel Defoe, *An Essay on the History and Reality of Apparitions* (1727), quoted in Peter Earle, *The World of Defoe* (New York: Atheneum, 1977), p. 40.

113 "flatter'd themselves with a perswasion": John Locke, *An Essay Concerning Human Understanding*, ed. Roger Woolhouse (London and New York: Penguin Books, 1997), p. 616.

113   a repudiation of mediators in all their manifestations: Ronald Knox, *Enthusiasm: A Chapter in the History of Religion* (Notre Dame, IN: University of Notre Dame Press, 1994), p. 585.

113   a "morbid distrust" of the intellect: Ibid., p. 586.

113   come into a "new order of being": Ibid., p. 3.

114   "Nonsense and Enthusiasm": Henry Fielding, *Joseph Andrews* (1742) (New York and London: Penguin Books, 1999), p. 113.

114   "ultrasupernaturalism": Knox, *Enthusiasm*, p. 4.

114   Fox . . . bummed around England: Reay, *The Quakers and the English Revolution*, p. 17.

115   the Restoration . . . after which the Quakers: Ibid., p. viii.

115   coiffed and bearded to look like Jesus: Knox, *Enthusiasm*, p. 163.

116   twenty thousand letters from pious women: Ibid., p. 262.

116   "an age of introverts": Ibid., p. 246.

117   giving priority to the God of power and majesty: Grace Davie, *Religion in Britain Since 1945* (Oxford: Blackwell, 1994), pp. 119–20.

117   "above nature, and beyond common human living": Quoted in David F. Noble, *A World Without Women* (Oxford and New York: Oxford University Press, 1992), p. 241.

CHAPTER ELEVEN

118   "'We also are religious'": Osborn, *Tertullian, First Theologian of the West*, p 1.

119   "Truth, is ever to be found in simplicity": Newton, Yahuda MS, 1 fol 14r, quoted in Manuel, *The Religion of Isaac Newton*, p. 43.

119   "the most basic unifying concept there is": John Wheeler, *At Home in the Universe* (New York: Springer-Verlag, 1996), p. 291.

119   "I know of no other reason": Ian Hacking, "Disunited Sciences," in Richard Q. Elvee, ed., *Nobel Conference XXV. The End of Science? Attack and Defense* (Lanham, MD: University Press of America, 1993), pp. 42–45.

120   He warned that the words "theory": Ibid., p. 61.

120   "does nothing in vain when less will serve": Isaac Newton, *Rules for Reasoning in Philosophy*, in I. Bernard Cohen and Richard S. Westfall, eds., *Philosophy of Newton: and Texts, Backgrounds and Commentaries* (New York and London: W. W. Norton & Company, 1995), p. 116.

121   "this vicious abundance of *Phrase*": Sprat, *The History of the Royal Society*, p. 92.

121   a list of "Common Notions": Harold R. Hutcheson, ed., Lord Herbert of Cherbury's *De Religione Laici* (New Haven: Yale University Press, 1944), p. 27.

122   "the signal gun of the deistic controversy": See John Toland, *Christianity Not Mysterious*, ed. Philip McGuiness, Alan Harrison, and Richard Kearney (Dublin: Lilliput Press, 1997), p. 237.

122   "from my cradle in the grossest superstition and idolatry": Ibid., p. 223.

122 "of that Sex that's least Proof against Flattery": Ibid., p. 42.

123 the "Naked Gospel": See John Herman Randall, *The Making of the Modern Mind* (Boston: Houghton Mifflin, 1940). Arthur Bury coined the term in 1690.

123 "unsociable in its morality": Ibid., p. 202.

124 But Franklin learned the hard way: Walter Isaacson, *Benjamin Franklin: An American Life* (New York: Simon & Schuster, 2002), p. 42.

124 "I began to suspect": Quoted in ibid., p. 46.

124 Franklin toyed with the idea: Ibid., p. 85.

125 "was, in fact, not merely an extrinsic ornament": Arthur O. Lovejoy, *The Great Chain of Being* (Cambridge, MA, and London: Harvard University Press, 1973), pp. 7–8.

125 "Nature's Simple Plan": See Chauncey B. Tinker, *Nature's Simple Plan* (Princeton and London, 1922).

125 "there was so little of the savage": See ibid., pp. 77–78.

126 "shamed education": Quoted in ibid., p. 78.

126 "It was not a question": Ibid., p. 29.

127 an "aristocratic" outlook: See Helge Kragh, *Dirac: A Scientific Biography* (Cambridge: Cambridge University Press, 1990), p. 278.

127 "Of all physicists, Dirac has the purest soul": Quoted in ibid., p. 72.

127 "It is not irrelevant to point out": Ibid., p. 231.

CHAPTER TWELVE

129 "My dear friend": Fraser, *Cromwell*, p. 267.

129 "Cromwell was about to ravage": Pascal, *Pensées*, pp. 136–37.

130 has tracked the rise of the legal profession: See William J. Bouwsma, *A Usable Past: Essays in European Cultural History* (Berkeley: University of California Press, 1990), pp. 129–53.

131 "Lawyers represented the growing assumption": Ibid., p. 142.

131 "By defining what was socially intolerable": Ibid., p. 143.

132 "In pamphlet after pamphlet": Wilfred Cantwell Smith, *The Meaning and End of Religion* (Minneapolis: Fortress Press, 1996), p. 40.

132 "a mountain torrent": Whale, *The Protestant Tradition*, p. 119.

133 a need to bring religion up to date: James Turner, *Without God, Without Creed* (Baltimore: Johns Hopkins University Press, 1986), p. 51.

133 "One might almost say": Smith, *The Meaning and End of Religion*, p. 19.

134 "We are moving toward": Robert Coles, ed., *Dietrich Bonhoeffer* (Maryknoll, NY: Orbis Books, 1998), p. 119.

134 "My resistance against everything": Quoted in Stanley Haverwas, *Performing the Faith* (Grand Rapids, MI: Brazos Press, 2004), p. 17.

134 "My suspicion and horror": Ibid.

134 "I have become doubtful": Coles, *Dietrich Bonhoeffer*, p. 121.

135 "Instead of discussing the character": Bernard Cooke, *The Distancing of God* (Minneapolis: Fortress Press, 1990), p. 238 (italics in the original).

136 "pious exaltations of the mind": John Dillenberger and Claude Welch, *Protestant Christianity* (New York: Prentice-Hall, 1988), p. 183.

136 "we have not trodden on the toes": Timothy J. Gorringe, *Karl Barth: Against Hegemony* (Oxford and New York: Oxford University Press, 1999), p. 61.

136 "precisely where human beings bolt and bar ": Ibid.

136 "for our culture against the uncultured": Ibid., p. 35.

136 "like drowning men": Ibid., p. 36.

137 "come out of polytheism": Ibid., p. 4.

137 "the Holy Spirit and Baptism": Trevor Hart, *Regarding Karl Barth* (Downers Grove, IL: Intervarsity Press, 1999), p. 146.

137 "including our religious pretensions": Dillenberger and Welch, *Protestant Christianity*, p. 256.

137 "The Bible tells us": Karl Barth, *The Word of God and the Word of Man* (New York: Pilgrim Press, 1986), p. 43.

CHAPTER THIRTEEN

139 "We adore the mysteries": Philip Melanchthon quoted in Jürgen Moltmann, *The Trinity and the Kingdom* (Minneapolis: Fortress Press, 1995), p. 1.

139 A contribution to this debate: See Fraser, *Cromwell*, p 17.

140 "men and visible helps": Ibid., p. 102.

140 "Since the Providence of God": Ibid., p. 275.

141 not been "the pleasure of God" Ibid., p. 292.

141 "Prints on his mad adventure's": Quoted in Christopher Hill, *God's Englishman: Oliver Cromwell and the English Revolution* (New York: HarperCollins, 1970), p. 234.

142 "hidden in the counsel of God": Calvin, *Institutes of the Christian Religion*, Vol. 1, p. 180.

142 "will learn, when it is too late": Ibid., p. 142.

142 "sluggish and groveling": Ibid., Vol. 2, p. 141.

142 Robert Nisbet . . . noted: See Robert Nisbet, *History of the Idea of Progress* (New York: Basic Books, 1980), p. 143.

143 a distancing move if ever there was one: Cooke, *The Distancing of God*, p. 238.

143 "nothing but the striving": Quoted in Jonathan I. Israel, *Radical Enlightenment* (Oxford and New York: Oxford University Press, 2001), p. 221.

144 "cynical audacity": Richard Price, *British Society 1680–1880* (Cambridge and New York: Cambridge University Press, 1999), p. 214.

144 "the sovereignty of popular will": Ibid.

144 "The God of order, governor of the universe": Jacob, *Newtonians and the English Revolution*, p. 96.

145 "God is no longer seen": Nisbet, *History of the Idea of Progress*, p. 129 (italics in the original).

145 A "natural and inexorable" pattern: Ibid.

146 "The ease with which Christ": Ibid., p. 136.

147 John Milton took to Arminianism: Hill, *Milton and the English Revolution*, p. 275.

148 "surrender to the disposal": Quoted in Maximillian E. Novak, *Daniel Defoe, Master of Fictions* (Oxford and New York: Oxford University Press, 2001), p. 221.

148 a "feeling" rather than a reality: Ibid., p. 223.

148 "intimations from heaven": Daniel Defoe, *Journal of the Plague Year* (1722) (London and New York: Penguin Classics, 2003), p. 11.

148 "flying from God": Ibid.

149 "the intimations which I thought I had": Ibid., p. 13.

149 "Upon the foot of all these observations": Ibid., p 169.

149 "The relative impotence": Ian Watt, *The Rise of the Novel* (Berkeley: University of California Press, 1957), p. 82.

150 "till Heaven beats the drum itself": Daniel Defoe, *Serious Reflections During the Life and Surprising Adventures of Robinson Crusoe* (1720), quoted in ibid., p. 82.

150 "scarce worth hanging": Daniel Defoe, *The Further Adventures of Robinson Crusoe* (1716) (Gillette, NJ: Wildside Press, 2004), p. 35.

150 a variation of the Reformation penchant: Michael McKeon, *The Origins of the English Novel 1600–1740* (Baltimore: Johns Hopkins University Press, 1988), p. 124.

150 "another sort of dispensation": Samuel Richardson, *Clarissa*, ed. Angus Ross (New York and London: Penguin Books, 2004), pp. 1495–96.

151 a "ridiculous doctrine": Quoted in ibid., p. 1497.

151 "In my humble opinion": Ibid.

152 "shaking off the yoake of the Beast": Quoted in Ernest Lee Tuveson, *Millennium and Utopia* (New York: Harper & Row, 1964), p. 19.

153 a "separate chosen people": Ernest Lee Tuveson, *Redeemer Nation* (Chicago: University of Chicago Press, 1968), p. 2.

153 "the Opening of a grand scene": Ibid., p. 102.

153 "Round thy broad fields": Ibid., p. 105.

153 "Follow him, in his strides": Ibid., pp. 117, 119.

153 "came into a kind of chemical": Ibid., p. 127.

154 "great things from our race": Ibid., p. 157.

154 "make good their redemption": Ibid., pp. 211, 213.

CHAPTER FOURTEEN

155   "If God wanted his tracks": John Updike, *Roger's Version* (New York: Fawcett Columbine, 1986), p. 174.

155   "Is it to be wondered": Quoted in Patrick Dillon Gin, *The Much Lamented Death of Madam Geneva* (Boston: Justin, Charles & Co., 2003), p. 3.

156   a "merchant political elite": See Robert Brenner, *Merchants and Revolution* (London and New York: Verso, 2003), p. 4.

156   "There is no place in the Town": Quoted in J. S. Bromley, ed., *The Rise of Great Britain and Russia, 1686–1725* (Cambridge and New York: Cambridge University Press, 1971), p. 282.

156   "sell the bear's skin before": Gin, *The Much Lamented Death of Madam Geneva*, p. 30.

157   "Take heed of computing": Quoted in Christopher Hill, *Some Intellectual Consequences of the English Revolution* (London: Phoenix Giant, 1980), p. 58.

157   "despaired of the meaning of history": Klaus Vondung, "Millenarianism, Hermeticism and the Search for a Universal Science," in Stephen A. McKnight, ed., *Science, Pseudo-Science and Utopianism in Early Modern Thought* (Columbia and London: University of Missouri Press, 1992), p. 120.

157   "proud speculative man": Ibid., p. x.

158   "the public good": Ibid., p. 117.

158   All talk of religion and politics was banned: Ibid., p. 176.

159   "that they might enjoy the benefits": Sprat, *The History of the Royal Society*, p. 91.

159   "they that contend for truth": Ibid., p. 92.

159   "heap up a mixt Mass": Ibid., p. 115.

159   "experiments of rarefaction, refraction": Ibid., p. 156.

159   But "the public": See James van Horn Melton, *The Rise of the Public in Enlightenment Europe* (Cambridge and New York: Cambridge University Press, 2001), p. 1.

160   supporters of a candidate were given: Ibid., p. 233.

161   "spread like leprosy": *The Diary of Samuel Pepys*, Vol. X, comp. and ed. Robert Latham (Berkeley: University of California Press, 2000), pp. 70–72.

161   helped to "deprivatise" society: Melton, *The Rise of the Public in Enlightenment Europe*, pp. 20-21.

161   the "rituals" of election politics: Ibid., p. 24.

161   "He has not been around": Ibid., p. 25.

162   a shift from "intensive" reading: Ibid., p. 89.

162   Skimming came into fashion: Ibid., p. 91.

162   The "desacrilisation and commodification": Ibid., p. 92.

163   It seems that theology needs science: *Newsweek*, July 20, 1998, p. 50.

163   "Theologians have grown grateful": Bertrand Russell, *The Scientific Outlook* (New York: W. W. Norton & Company, 1931), p. 110.

164   "we may think of the finger of God": James Jeans, *The Universe Around Us* (New York: The Macmillan Company, 1929), p. 317.

164   a "thought in the mind of the Creator": Helge Kragh, *Cosmology and Controversy* (Princeton: Princeton University Press, 1996), p. 42.

164   Lemaître's letter to *Nature*: Quoted in ibid., p. 48.

164   "much more abstruse": *The Literary Digest*, 115, March 11, 1933, p. 23.

165   These efforts regularly ended in disaster: Simon Singh, *Big Bang: The Origin of the Universe* (New York and London: Harper Perennial, 2004), p. 45.

165   "If you're religious, it's like seeing": Quoted in Karen C. Fox, *The Big Bang Theory* (New York: John Wiley & Sons, 2002), pp. 113–14.

165   Milne unblushingly linked his theories: See Kragh, *Cosmology and Controversy*, p. 66.

165   "If evolution consists in": Ibid., p. 259.

167   has argued for setting up a world tribunal: Robert Drinan, *Can God and Caesar Coexist? Balancing Religious Freedom and International Law* (New Haven and London: Yale University Press, 2004), p. 42.

167   "In a sense, the Catholic Church": Ibid., p. 108.

168   A surreptitious version of that devotion: Donald Cupitt, *The Worlds of Science and Religion* (New York: Hawthorn Books, 1976), p. 105.

169   "In a curious way, Christianity seems to *need*": Ibid., p. 20 (italics in the original).

CHAPTER FIFTEEN

170   "There cannot be a personal God": Cyril Connolly, *The Unquiet Grave* (New York: Persea Books, 1981), p. 16.

170   to "cut God down to size": Christopher Hill, *Change and Continuity in Seventeenth Century England* (New Haven and London: Yale University Press, 1991), p. 114.

171   "no ordinance, human or from heaven": John Milton, *Complete Prose Works,* Vol. I, quoted in ibid.

171   "we ought to be very tender": Quoted in ibid., pp. 113–14.

173   "There is no subject on which there is so much difference": James Gustafson, "Alternative Conceptions of God," in Thomas F. Tracy, ed., *The God Who Acts* (University Park, PA: Pennsylvania State University Press, 1994), p. 68.

174   One of the major shifts in Protestant thought: Ibid., p. 71.

174   "evolving a more correct idea": Malachi Martin, *The Jesuits* (New Yorki: Simon & Schuster, 1987), p. 19.

174   "If it were not for that knowledge": Gustafson, "Alternative Conceptions of God," p. 71.

175   "It may be that the task of ethics": James Gustafson, *Ethics from a Theocentric Perspective* (Chicago: Chicago University Press, 1981), p. 113.

175 "most Anglican bishops are content": David Edwards, *Tradition and Truth* (London: Hodder & Stoughton, 1989), p. 266.

176 "although accustomed to a monarchy": Ibid.

176 leads to a sort of despair: Ibid., p. 267.

177 "where things are full of nothing but cursing and blasphemy": Aaron Lichtenstein, *Henry More: The Rational Theology of a Cambridge Platonist* (Cambridge, MA: Harvard University Press, 1967), p. 6.

177 "a small compendious transcript": Ibid., p. 100.

177 More got a letter from René Descartes: See Placher, *The Domestication of Transcendence*, pp. 131–32.

177 "the whole machine of the world": Ibid., pp. 132–33.

178 "exists in every little seed": Ibid., p. 133.

178 to "identify and kill": Ibid., p. 134.

178 "What sophistical knots and Nooses": Henry More, *Mystery of Godliness*, V, xiv, 7, quoted in Lichtenstein, *Henry More*, p. 123.

179 "It stares back at you": Karl Barth and Emil Brunner, *Natural Theology* (Eugene, OR: Wipf & Stock, 2002), p. 76.

180 "were it vastly more superstitious": John Henry Newman, *Parochial and Plain Sermons*, Vol. 1 (London: Longmans Green & Co., 1891), p. 230.

CHAPTER SIXTEEN

181 "When Christianity was not doctrine": Søren Kierkegaard, *Papers and Journals: A Selection,* ed. Howard Hong (London and New York: Penguin Books, 1996), p. 616.

184 "the orthodox Jew had no physical or ritualistic crutch": Norman Cantor, *Antiquity* (New York: HarperCollins, 2003), p. 91.

186 as "a man is to his own jugular vein": John MacQuarrie, *Mediators Between Human and Divine* (New York: Continuum Press, 1999), p. 125.

186 thrown out of a second-story window: See Robert Spencer, *Islam Unveiled* (San Francisco: Encounter Books, 2002), p. 4.

187 An "anti-intellectual rage": Ibid., p. 125.

187 "God's hands are chained": Ibid., pp. 126–27.

188 "threatens to become explosive": Ibid., p. 129.

188 beliefs basic to the "theological culture": Fred Hoyle, *The Intelligent Universe* (New York: Book Sales Reprint, 1988), p. 71.

189 "Indeed, I think that every Christian sect": Quoted in Michael Buckley, *At the Origins of Modern Atheism* (New Haven and London: Yale University Press, 1987), p. 377.

190 "One was informed about God": Ibid., p. 346.

191 "just as the northern tribes": Ibid., p. 360.

191 "To insist that God is personal": Ibid., p. 363.

CHAPTER SEVENTEEN

192 "Not only did Luther love superlatives": Joseph Lortz, "Luther and the Reformation," in Norman F. Cantor and Michael S. Werthman, eds., *Renaissance, Reformation and Absolutism, 1450–1650* (New York: Thomas Y. Crowell Company, 1972), p. 114.

192 "This little word 'why'": Fyodor Dostoyevsky, *The Devils*, trans. by David Margarshack (London and New York: Penguin Books, 2004), p. 184.

193 "I can believe anything": See Roger Scruton, *Modern Philosophy: A Survey* (London: Sinclair-Stevenson, 1994), pp. 245–46.

194 Once, at a dinner party: Jacob Bronowski, *The Origins of Knowledge and Imagination* (New Haven: Yale University Press, 1978), pp. 78–79.

194 "If I *must* get to heaven": Johnson, *The Theology of John Donne*, p. 8.

195 who "holds the universe in his hand": Osborn, *Tertullian, First Theologian of the West*, p. 11.

195 "God is wholly other": Ibid., p. 62.

196 the dualists implicitly "blaspheme": Tillich, *A History of Christian Thought*, pp. 36–37.

197 "breeding maggots out of the dung": Updike, *Roger's Version*, p. 176.

197 "the one we care about in this divinity school": Ibid., p. 22.

197 "has some of the hardheartedness": James Plath, ed., *Conversations with John Updike* (Jackson, MS: University Press of Mississippi, 1994), p. 93.

197 "A god who is not God": Ibid., p. 33.

197 "God can't hide any more": Updike, *Roger's Version*, p. 21.

198 of "divine purpose, more or less": Ibid., p. 179.

198 "two conglomerations of vertices": Ibid., p. 244.

198 "a certain optical alignment": Ibid., p. 250.

198 "was perfecting his imitation of that Heavenly Presider": Ibid., p. 128.

199 "would have a certain otherworldliness": Quoted by Darrell Jodock, "What Is Goodness?" in James Yerkes, ed., *John Updike and Religion* (Grand Rapids, MI, and Cambridge: William B. Eerdmans, 1999), p. 121.

199 "His cannons now seemed to roll about": Ozment, *Protestants: The Birth of a Revolution*, p. 198.

200 "The question that drove Martin Luther": Ibid., p. 200.

200 the experience filled him with disgust: Marius, *Martin Luther*, pp. 73–75.

200 Staupitz . . . told Luther he lacked a developed sense: Ibid., pp. 75–76.

201 "Luther could not speak of God's love": Ibid., pp. 77–78.

202 "The thinker without a paradox": Kierkegaard, *Philosophical Fragments*, chapter 3, quoted in ibid., p. 398.

202 "he would become ludicrous": Kierkegaard, *Papers and Journals: A Selection*, pp. 615–16.

203 "ironic science": John Horgan, *The End of Science* (New York and Reading, MA: Helix Books, Addison-Wesley Publishing Co., 1998), p. 4.

204 "At its best, ironic science": ibid., p. 8.

205 "An appreciation of the transience of theories": Cyril Domb in Richardson et al., *Science and the Spiritual Quest*, p. 66.

CHAPTER EIGHTEEN

206 "Irony is an abnormal growth": Kierkegaard in Dru, *The Soul of Kierkegaard*, p. 58.

206 "Deepest down in the heart of piety": Kierkegaard, *Philosophical Fragments*, p. 134.

207 an opinion survey conducted in England in 2004: See Anthony King, "Britons' Belief in God Vanishing as Religion Is Replaced by Apathy," *Daily Telegraph* (London), December 27, 2004, p. 16.

208 "had bestowed upon the English people": Donald Cupitt, *Christ and the Hiddenness of God* (London: SCM Press, 1985), p. 5.

208 "against the enslavement of men": Nigel Leaves, *Odyssey on the Sea of Faith* (Santa Rosa, CA: Polebridge Press, 2004), p. 19.

209 was being described as a "fallen angel": Ibid.

209 "In fables and fairy tales": Søren Kierkegaard, *Concluding Unscientific Postscript*, ed. Howard V. Hong and Edna H. Hong (Princeton: Princeton University Press, 1992), p. 138.

210 a kind of "religious dandyism": Donald Cupitt, *Taking Leave of God* (London: SCM Press, 2001), p. xvi.

211 It "systematically and in principle refuses": Ibid., p. 42.

211 "a hell of a row": Leaves, *Odyssey on the Sea of Faith*, p. 30.

212 "The sense of the Divine vanishes": Iris Murdoch, *Metaphysics as a Guide to Morals* (London: Penguin Books, 1995), p. 56.

213 "for it searches the heart": Cupitt, *Taking Leave of God*, p. 93.

213 "Protestant success against medieval religion": Ozment, *The Age of Reform, 1250–1550*, p. 438.

214 "He never fully lets go": Leaves, *Odyssey on the Sea of Faith*, p. 11.

214 the Kingdom of Anxiety: William Hubben, *Dostoevsky, Kierkegaard, Nietzsche and Kafka* (New York: Collier Books, 1962), p. 34.

214 But six years later, he announced: See Cupitt, *Taking Leave of God*, p. 37.

CHAPTER NINETEEN

216 "So creedless is the American Religion": Harold Bloom, *The American Religion* (New York and London: Simon & Schuster, 1992), p. 28.

216 "For man has not deduced": Miguel de Unamuno, *The Tragic Sense of Life*, trans. J. E. Crawford Flitch (New York: Dover Publications, 1954), p. 156.

217 sees the emergence of a science of religion: Peter Harrison, *Religion and*

the Religions in the English Enlightenment (Cambridge and New York: Cambridge University Press, 1990), p. 2.

217 "Most of those who lived": Christopher Haigh, *English Reformations* (Oxford: Clarendon Press, 1994), p. 21.

217 the "loathsome contempt, hatred and disdain": Quoted in Hill, *The World Turned Upside Down*, p. 23.

217 "that she did not care a pin nor a fart": Ibid., p. 24.

218 a time of "glorious flux": Ibid.

218 "the humour of being witty": Quoted in Samuel Mintz, *The Hunting of Leviathan* (Cambridge: Cambridge University Press, 1962), p. 136.

219 "Men of little Genius": Quoted in D. Judson Milburn, *The Age of Wit, 1650–1750* (New York: Macmillan, 1966), p. 276.

219 "Chuses the Councel and Officers": William Stukeley, *Family Memoirs,* Vol. I (1780), quoted in James E. Force, "Hume and the Relation of Science to Religion," in John W. Yolton, ed., *Philosophy, Religion and Science in the Seventeenth and Eighteenth Centuries* (Rochester, NY: University of Rochester Press, 1981), p. 229.

220 Abraham's ambiguity as an actual character: See Bruce Feiler, *Abraham, A Journey to the Heart of Three Faiths* (New York: HarperCollins, 2002), p. 21.

220 "alternative spiritual systems": James A. Jerrick, *The Making of the New Spirituality* (Downers Grove, IL: Intervarsity Press, 2003), p. 17.

222 "as strikingly similar to the society": Alan Wolfe, *The Transformation of American Religion* (New York: Free Press, 2003), p. 3.

222 Wolfe calls this "God Lite": Ibid., p. 157.

223 "even if I feel mad at him": Ibid., p. 160.

223 "unknowable but powerful and authoritative": Ibid., p. 161.

223 "do not only deny *them*": Sir Thomas Browne, *Religio Medici* (1643) (Cambridge: Cambridge University Press, 1955), p. 40.

223 "symbol of how much of the endless": Basil Willey, *The Seventeenth Century Background* (New York: Columbia University Press, 1934), p. 55.

224 "We don't feel holy": Quoted in Wolfe, *The Transformation of American Religion*, p. 162.

224 "America's God has been domesticated": Ibid., p. 165.

225 "a whole eternity of perfect obedience": Quoted in Ann Douglas, *The Feminization of American Culture* (New York: Avon Books, 1977), p. 143.

226 "Crushing, humiliating as it may appear": Ibid., p. 146.

226 a "creature of feeling": Ibid., p. 151.

226 Modern man . . . has "an easy conscience": Reinhold Niebuhr, *The Nature and Destiny of Man.* Vol. I: *Human Nature* (Englewood Cliffs, NJ: Prentice-Hall, 1964), p. 1.

227 "These people believe": Wolfe, *The Transformation of American Religion*, p. 72.

228 "cafeteria religion": Ingolf Dalferth, " 'I Determine What God Is!' Theology in an Age of Cafeteria Religion," *Theology Today*, vol. 57, no.1 (April 2000), p. 5.

228 "scraps from the world's religions": Ibid., p. 6.

228 "esoteric cults of mystery and meditation": Ibid., p. 23.

229 "apparently under the impression": Dorothy Sayers, *The Whimsical Christian* (New York: Macmillan, 1978), p. 23.

229 "were looked upon as astonishing": Ibid., pp. 25–26.

229 a religion of "conscience and decision": Dillenberger and Welch, *Protestant Christianity*, p. 150.

230 "something like the ancient misalliance": Philip J. Lee, *Against the Protestant Gnostics* (New York and Oxford: Oxford University Press, 1987), p. xiv.

CHAPTER TWENTY

231 "What soul ever perished": James Como, ed., *C. S. Lewis at the Breakfast Table* (New York: Harcourt Brace Jovanovich, 1979), p. 13.

232 Readers "understand and appreciate": Wade Roof, *Spiritual Marketplace* (Princeton: Princeton University Press, 2001), p. 201.

233 God's "anxiety": Miles, *God, A Biography*, p. 6.

233 is fond of stories of deception and fraud: Ibid., p. 32.

233 "What polytheism would allow": Ibid., p. 33.

234 he is unemotional, unruffled: Ibid., p. 43.

235 Possibly there occurs here a synthesis of gods: Ibid., p. 72.

235 which is quite a high-handed way of carrying on: Ibid., p. 62.

235 becomes a "friend of the family": Ibid., p. 67.

236 "two manner of people": Genesis 25:23.

236 "managing the pregnancies one by one": Miles, *God, A Biography*, p. 68.

236 "God is like a novelist": Ibid., p. 86.

237 "the profound originality": Ibid., p. 133.

238 "What struck me about those conversations": Quoted in the *New York Times Magazine*, December 7, 1997, p. 58.

239 Ward blames science for the popular fallacy: Keith Ward, *God: A Guide for the Perplexed* (Oxford: OneWorld Publications, 2002), pp. 50–51.

240 faith is not something added: Ibid., p. 60.

241 that the false historical claims showed that Christianity and history: Charlotte Allen, *The Human Christ* (New York: Free Press, 1998), p. 238.

242 He absorbed the theory: See ibid., p. 240.

242 in an excellent summary of this troubled period: Ibid., p. 247.

243 Even these supposedly hard facts are only provisional: Ibid., p. 249.

243 of the images of "Jesus" held dear by Americans: Stephen Prothero, *American Jesus: How the Son of God Became a National Icon* (New York: Farrar, Straus & Giroux, 2003), p. 11.

243 and finally an icon and a trademark: Ibid., p. 12.

243 faith became a matter of individual choice: Ibid., p. 46.

243 the "maniac ravings" of Calvin: Ibid., p. 52.

244 "making humans more divine": Ibid.

244 the "noxious exaggeration": Quoted in ibid.

244 "You name the good Jesus": Quoted in ibid., p. 53.

244 "they seem to focus more of their devotion": Ibid., p. 59.

244 "the praying wing of the Woodstock Nation": Ibid., p. 125.

245 ran a Jesus Revolution cover: *Time* magazine, June 21, 1971.

245 a "common cultural coin": Prothero, *American Jesus*, p. 300.

CHAPTER TWENTY-ONE

246 "We do not imagine God to be lawless": Calvin, *Institutes of the Christian Religion*, Vol. 2, p. 237.

246 "A God you could prove": Updike, *Roger's Version*, p. 24.

246 "Who are you to tell God": Thomas Levenson, *Einstein in Berlin* (New York and London: Bantam Books, 2003), p. 343.

247 "splendour, sobriety and coherence": C. S. Lewis, *The Discarded Image* (Cambridge and New York: Cambridge University Press, 1964), p. 216.

248 "If it had an aesthetic fault": Ibid., p. 121.

248 hedge it within a tidy plot of cultivated land: Polkinghorne, *The Way the World Is*, p. 15.

248 "Big and old and dark and cold": Michael Poole, "Big and Old and Dark and Cold," in Russell Stannard, ed., *God for the Twenty-first Century* (Philadelphia: Templeton Foundation Press, 2000), pp. 15–17.

249 "pocket universes": Leonard Sullivan, *The Cosmic Landscape* (New York and Boston: Little, Brown, 2005), p. 14.

249 "Not even the three dimensions": Ibid., p. 20.

249 "The question 'Why is the universe'": See ibid., pp. 13–14.

249 "Every person's concept of God": Quoted in Robert L. Hermann, ed., *God, Science and Humility* (Philadelphia and London: Templeton Foundation Press, 2000), p. vii.

250 "principle of mediocrity": Owen Gingerich, "The Arrogance of Mediocrity," in ibid., p. 118.

250 it had to be; after all, it was ours: Ibid., p. 122.

250 "The scale of theological thinking": John Polkinghorne, *Belief in God in an Age of Science* (New Haven and London: Yale University Press, 1998), p. 87.

250 that God is infinitely greater than anything: Gingerich, "The Arrogance of Mediocrity," p. 128.

251 what it might be "reasonable to suppose": Polkinghorne, *The Way the World Is*, pp. ix–x.

251 "Our powers of reasonable prevision": Ibid., p. x.

252 "Perhaps the desire to make God": Ibid., p. 15.

252 has moved . . . to a more crucial focus: John Polkinghorne, *Faith, Science and Understanding* (New Haven and London: Yale University Press, 2000), p. 105.

253 the apparently simple case of billiard balls colliding: John Polkinghorne, *Science and Providence* (Boston: New Science Library/Shambala, 1989), p. 28.

253 an "intrinsic openness": Ibid., p. 29.

254 "The universe may not look": Ibid., p. 30.

254 "creative imagination": Keith Ward, *Divine Action* (London: Collins, 1990), p. 69.

254 Schrödinger's choice of animal: Erwin Schrödinger, *Autobiographical Sketches* (Cambridge and New York: Cambridge University Press, 1989), p. 176.

254 they have to overcome the tremendous repulsive force: Owen Gingerich, "The Universe as a Theater for God's Action," *Theology Today*, vol. 55, no. 3 (1998), p. 314.

255 Poincaré . . . came to the conclusion: Ibid.

255 Ward almost paraphrases Newton: See Ward, *Divine Action*, p. 69.

256 "do not state the truth": Nancy Cartwright, *How the Laws of Physics Lie* (Oxford and New York: Oxford University Press, 1983), p. 3.

256 "God may have written": Ibid., p. 49.

256 The difference . . . "is almost theological": Ibid., p. 19.

257 the tiny "blur of uncertainty": Gingerich, "The Universe as a Theater for God's Action," p. 315.

257 "would have been as sour as battery acid": Ibid., p. 312.

257 "But the theist could nonetheless argue": Ibid., p. 315.

257 One leading sponsor of the theory: See Nancey Murphy, "Divine Action in the Natural Order: Buridan's Ass and Schrödinger's Cat," in Robert John Russell, Nancey Murphy, and Arthur R. Peacocke, eds., *Chaos and Complexity: Scientific Perspectives on Divine Action* (Vatican City State: Vatican Observatory and Berkeley: Center for Theology and Natural Sciences, 1996), pp. 325–57.

258 that just to steer into our planet: Nicholas Saunders, "Does God Cheat at Dice? Divine Action and Quantum Possibilities," *Zygon*, vol. 35, no. 3 (September 2000), p. 540.

258 "dislikes the idea of God scrabbling around": Ward, *Divine Action*, p. 127.

259 "I do not find": John Polkinghorne, *The Faith of a Physicist: Reflections of a Bottom-Up Thinker* (Minneapolis: Fortress Press, 1996), p. 1.

260 He is "omnitemporal": Grace Jantzen, *God's World, God's Body* (Philadelphia: Westminster Press, 1984), p. 41.

260 drops one of his biggest theological percussion bombs: Polkinghorne, *Science and Providence*, p. 79.

CHAPTER TWENTY-TWO

261   "One feels inclined to say": Sigmund Freud, *Civilization and Its Discontents* (New York and London: W. W. Norton & Company, 1961), p. 25.

261   His workings in history: John Polkinghorne, *Science and the Trinity* (New Haven and London: Yale University Press, 2004), p. 54.

262   When Abraham pleads with God: See Polkinghorne, *Science and Providence*, p. 79.

262   is "precarious": Polkinghorne, *Serious Talk*, p. 85.

262   the "cloudy unpredictabilities": Ibid., p. 84.

263   "just a glimmer (no more)": Polkinghorne, *Faith, Science and Understanding*, pp. 96–97.

263   "information input": John Polkinghorne, *Reason and Reality: The Relationship Between Science and Theology* (Philadelphia: Trinity Press International, 1991), p. 45.

263   "the director of the great cosmic" Polkinghorne, *Faith, Science and Understanding*, p. 125.

263   "improviser of unsurpassed ingenuity": Polkinghorne, *Reason and Reality*, p. 46.

263   "Whether acknowledged or not": Polkinghorne, *Science and the Trinity*, p. 104.

263   "without acknowledging": Hans Christian von Bayer, *Information, the New Language of Science* (Cambridge, MA: Harvard University Press, 2004), p. 227.

264   "Every item in the physical world": Wheeler, *At Home in the Universe*, p. 296.

264   "a kind of information theory": Bloom, *The American Religion*, p. 30.

264   an "entity among other entities": Polkinghorne, *Reason and Reality*, p. 56.

265   "and the theologians who had deposited": Buckley, *At the Origins of Modern Atheism*, pp. 358–59.

266   stands the possibly irrational force of will: Tillich, *A History of Christian Thought*, pp. 191–92.

266   the "sinister" image of God: David Penchansky, *What Rough Beast? Images of God in the Hebrew Bible* (Louisville, KY: Westminster John Knox Press, 1999), p. 1.

266   a profoundly complex inquiry into the riddle: David Penchansky, *The Betrayal of God: Ideological Conflict in Job* (Louisville, KY: Westminster John Knox Press, 1990), p. 75.

266   "but rather an unrestrained shriek": Ibid., p. 83.

266   God who does not . . . seem to be in full control: Polkinghorne, *Faith, Science and Understanding*, p. 126.

267   a residue of Canaanite polytheism: Ibid., p. 75.

267   "some divine responsibility for human evil": Ibid.

267   Integrity is more human than divine: Ibid., p. 83.

268 "loyalty to one's own experience": Ibid., p. 50.

268 "One might even wonder": Ibid., p. 52.

268 the cult of ancient Yahwism: Ibid., pp. 56–57.

268 any reliable "relationship of obligation": Ibid., p. 75.

269 a massive, exuberant upsurge: "Spirituality in America: Our Faith Today," *Newsweek*, September 5, 2005, pp. 46–65.

269 "Israelite pain was driven inwards": Penchansky, *The Betrayal of God*, p. 84.

270 "Tyger, Tyger, burning bright": William Blake, quoted in Penchansky, *What Rough Beast?*, p. 151.

270 "perhaps even an evil God": Ibid., p. 4.

270 "might provide a structure": Ibid., p. 94.

271 "A permanent peace between science and religion": Herbert Spencer, *First Principles*, quoted in Bernard J. Verkamp, *The Senses of Mystery* (New York: University of Scranton Press, 1997), p. 112.

Index